$ 6.99

"A dazzling achiev
—*New York Times Bo*

D0038321

The Human Age

The World Shaped by Us

DIANE ACKERMAN

author of **THE ZOOKEEPER'S WIFE**

Praise for
THE HUMAN AGE

"Diane Ackerman's vivid writing, inexhaustible stock of insights, and unquenchable optimism have established her as a national treasure, and as one of our great authors. If you've read any of her previous books, you already know why you'll love this latest one. If you haven't read her previous books, you're now about to become addicted to Diane Ackerman." —Jared Diamond,
professor of geography at UCLA and Pulitzer Prize–winning
author of *Guns, Germs, and Steel* and *Collapse*

"Ackerman has established herself over the last quarter of a century as one of our most adventurous, charismatic and engrossing public science writers. . . . [S]he has demonstrated a rare versatility, a contagious curiosity and a gift for painting quick, memorable tableaus drawn from research across a panoply of disciplines. . . . *The Human Age* is a dazzling achievement: immensely readable, lively, polymathic, audacious." —Rob Nixon, *New York Times Book Review*

"Fascinating. . . . Ackerman offers a cross-cultural tour of human ingenuity. . . . Her words invite us to feel the hope she feels." —Barbara J. King, *Washington Post*

"What a hopeful book! We can feel many things at the start of the Anthropocene, this new geological age that we have brought on—or brought down on ourselves. Diane Ackerman, for one, is optimistic, even cheerful, singing the praises of those individuals among us who are helping to find the way forward. And she writes with brilliance, zest, and high style. We need to hear this voice of human affirmation. It's important. It matters. I read *The Human Age* and thought, Yes! This is the way to look ahead." —Jonathan Weiner,
Pulitzer Prize–winning author of
The Beak of the Finch and *Long for This World*

"Diane Ackerman guides us into the Anthropocene Era as an honest storyteller making the hard news ahead of us not something beyond ourselves, but of our own making. *The Human Age* allows us to consider whether or not we will accept destruction or restoration as our legacy. . . . With graceful intelligence, Ackerman calls for an enlightened guardianship for the planet. While living inside the pages of this book, I kept asking myself, 'Can we love this beautiful, broken world enough to change?' I cannot imagine a richer text of image and insight."

—Terry Tempest Williams,
author of *When Women Were Birds*

"A humdinger of a book. . . . Ackerman is optimistic, even exhilarated, and frequently giddy about the future of humanity."

—Jon Christensen, *San Francisco Chronicle*

"Naturalist Ackerman offers a prismatic vision of the genius and devastation that is mankind's legacy."

—Sarah Meyer, *O, The Oprah Magazine*

"[A] panoramic investigation of vertical ocean gardening, geo-friendly architecture, Wakodahatchee Wetlands, the 'bounty' that grows on planted urban walls, the coming age of regenerative medicine. . . . [Ackerman] is boundlessly, indefatigably curious . . . inherently hopeful as she plumbs the currents of now."

—Beth Kephart, *Chicago Tribune*

"Along with the utterly compelling examples in this book, *The Human Age*'s greatest strengths are in the beauty of Ackerman's language and the power of her metaphors. Whether she is writing about Polynesian snails whose 'interiors belong in a church designed by Gaudí,' or the astounding Quai Branly Museum in Paris in which 'multi-textured meadows climb the thirteen-thousand-square-foot facade of the building, more than half of which is alive,' her penetrating insight is a joy to behold." —Michael Zimmerman, *Bellingham Herald*

"[W]ith a poet's soul and a journalist's precision . . . [Ackerman's] examinations of overpopulation, the energy crisis, and our dependence on technology make *The Human Age* both foreboding and inspirational." —Kyle Anderson, *Entertainment Weekly*

"[A] thought-provoking analysis of our connection to the earth. . . . [A] lens that magnifies and clarifies the fascinating, far-reaching effects humans have had on our planet and ourselves."
—Lee E. Cart, *Shelf Awareness*

"Ackerman does not hide her optimism behind academic arguments. She moves quickly, a giddy tour guide shepherding readers around a busy world, sometimes delighted, sometimes fearful, always lively."
—Bill Streever, *Dallas Morning News*

"There is no writer now, perhaps ever, who is able to convey the wonder and magic of science with poetry comparable to Diane Ackerman."—Rick Searle, *Institute for Ethics & Emerging Technologies*

"She has great empathy for all her subjects, human and otherwise, allowing us to feel the excitement and joy they have in their pursuits. Above all she is a gorgeous writer of prose."
—Amanda Martin, *Neworld Review*

"It's not only great writing by a master prose stylist, it's also a wake up call." —Dan Bloom, *San Diego Jewish World*

"Exquisite and startling." —Tim Flannery, *Harper's*

DIANE ACKERMAN

THE HUMAN AGE

THE WORLD SHAPED BY US

W. W. NORTON & COMPANY

Independent Publishers Since 1923

NEW YORK LONDON

For information about permission to reproduce selections from this book,
write to Permissions, W. W. Norton & Company, Inc.,
500 Fifth Avenue, New York, NY 10110

For information about special discounts for bulk purchases, please contact
W. W. Norton Special Sales at specialsales@wwnorton.com or 800-233-4830

Manufacturing by RR Donnelley Westford
Book design by Jam design
Production manager: Julia Druskin

Library of Congress Cataloging-in-Publication Data

Ackerman, Diane, 1948–
The human age : the world shaped by us / Diane Ackerman. — First Edition.
    pages cm
Includes biographical references and index.
ISBN 978-0-393-24074-0 (hardcover : alk. paper)
1. Human ecology. 2. Civilization—History. 3. Human beings—History. 4. Nature—Effect
of human beings on. I. Title.
GF13.A35 2014
304.2 —dc23

                            2014027691

ISBN 978-0-393-35164-4 pbk.

W. W. Norton & Company, Inc.
500 Fifth Avenue, New York, N.Y. 10110
www.wwnorton.com

W. W. Norton & Company Ltd.
Castle House, 75/76 Wells Street, London W1T 3QT

1 2 3 4 5 6 7 8 9 0

CONTENTS

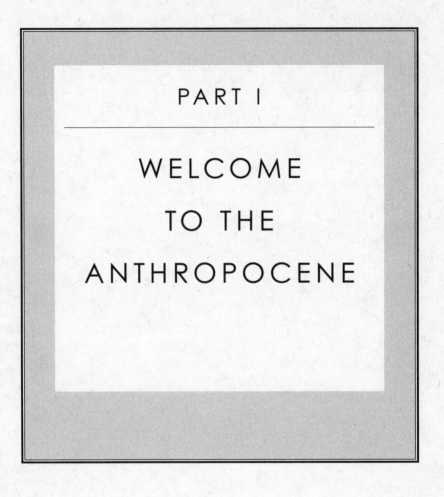

PART I

WELCOME TO THE ANTHROPOCENE

APPS FOR APES

On a blue-sky day at the Toronto zoo, flocks of children squired by teachers and parents mingle excitedly between exhibits. Some kids pull out cell phones and send texts or snap pictures with the easy camaraderie of wired life. Clustered noisily along a large domed habitat that's been designed to look like a multistoried Indonesian forest complete with tree nests and meandering stream, they watch two orangutan moms and young weaving fluently through a maze of thick, flat vines, which in reality are fire hoses. Orangutans are the swivel-hipped aerialists of the ape world, with ankle-length arms built for sky-walking, opposable thumbs *and* big toes, swervy knees, and bowed ankles. As a result they can twist into almost any angle or pose. In amazement I watch a young female swing smoothly from vine to vine, then grab two with wide-spread hands and feet, flatten her hips, rotate her wrists, and hang still as an orange kite snagged in the treetops.

Even with knuckle-walking far behind us, we sometimes feel the urge to brachiate in that way, hand over hand on a playground's "monkey bars." Yet we're stiff-jointed and feeble by comparison. We may share 97 percent of our genes with orangutans, but they remain the ginger-haired tree-dancers and we the chatterbox ground-dwellers. In the wild, orangutans spend most of their lives aloft, maneuvering with pendulous grace, as they pursue mainly solitary lives, except during childrearing. Moms raise one kid every six to eight years, doting on their young and teaching them the ways of the forest, where edible fruits abound but must be safely judged—and some aren't easy to peel or crack open because the rinds are either tough or spiked like medieval weaponry.

One of the orang moms swoops down to the ground as if on an invisible slide, picks up a stick, and fishes around inside a tree trunk until she snares edibles that she coaxes up and eats. The once-raucous students grow quietly transfixed as they peer at her skillful tool-using, especially the way she downs the morsels like eating peas off a knife.

Beyond the glade, well away from the crowds, I find a long-haired seven-year-old boy staring intently at an iPad and tapping the screen with one finger, which unlooses the pocket-sized roar of a lion followed by the buzzy honk of a flamingo. He glances at me with big brown almond-shaped eyes under a shag of thin auburn hair.

If my mane of black hair, frizzed wide in the heat, amuses him, he doesn't laugh. After holding my gaze for the sheerest moment, he turns back to his way-more-interesting iPad, gripping it with both hands, then with hands and naked feet. Surprisingly clean feet, I must say, and the largest hands I've ever seen on a seven-year-old. My whole hand would fit into his palm.

But that's not unusual for a Sumatran orangutan, and Budi, whose name means "Wise One" in Indonesian, is growing quickly and starting to show signs of puberty: the peach-fuzz beginnings of a mustache and beard, and the billow of what will one day be a majestic double chin that puffs and vibrates when, as a two-hundred-pound adult, he gulp-warble-croons his operatic "long call." There's no sign yet of the giant cheek pads between the eyes and ears that will frame his face, acting as a megaphone to shoot his long calls half a mile through dense canopies.

His companion, Matt Berridge, is a tall, slender, forty-something, dark-haired man, holding the iPad near the bars so that Budi can play with it but not drag it off and deconstruct it. The zoo's main orangutan keeper, Matt is the father of two young sons, both iPad devotees. Ape boys will be ape boys, after all.

The Apps for Apes program is sponsored by Orangutan Outreach, an international effort to help wild orangutans, whose bands are dwindling, and improve the lives of those in captivity around the world, by providing mental enrichment and more stimulating habitats. Nourishing the mind is a high priority because these great apes are about as smart as human three- to four-year-olds, and just as inquisitive. Clever tool-users, they wield sticks for many purposes, from batting down fruit to fishing for ants and termites. They fashion leaf gloves to protect their hands while eating thorny fruit or climbing over prickly vines. Day-dwellers, they fold a fresh mattress of leaves in the canopy before sunset each day. They lift leaf parasols overhead to shelter from extreme sun and fold leaf hats and roofs to keep off the rain. For drinking water, in a pinch, they chew and wad up leaves to make a sponge, then dip it into rain-filled plants. Before crossing a stream, they'll measure the water depth with a branch. They build dynamic mental maps of all the food trees in their leaf-cloud canopies.

And like their human counterparts, orangs enjoy playing with iPads. But they're not addicted to them. They're just not as enthralled by technology as we are.

"Like, I have a seven-year-old son," Matt tells me. "*He's* on it all the time. Not Budi."

This kid likes the luminous screen, but he wouldn't sit for cramped hours just staring at it.

"How are *we* so enamored of this thing that's so unnatural and takes you away from everything?" Matt asks. "In one way, you'd like to have your own kids occupied at times, but when you see that the orangs are never going to get obsessed with it, knowing their huge intelligence, it gets me thinking: How smart are *we* to spend all this time staring at this *thing*? Like, even myself, you know, I don't *test* my memory anymore. I go . . . *dee deedle dee dee.*" He demo-types on the screen. "I've become almost totally dependent on these machines. So am I weakening my brain?"

"Strawberry," a woman's voice says as Budi taps the strawberry on the screen. "Strawberry," she repeats when he finds a match. Matt rewards him with tidbits of fresh strawberries, apples, and pears. The lush tropical rainforests of Sumatra offer a cornucopia of hundreds of exotic fruits, the orangutan's favorite fare.

Another app—of pooling water—fascinates Budi. It looks like water, ripples like water, and when he touches it, it plashes and burbles. But it doesn't feel wet. And when he lifts his fingers to his nose, he doesn't smell water. From his sensory perspective, it's strange. Not as strange, though, as interacting with humans and other orangutans via Skype.

The first time Budi saw Richard Zimmerman, the director of Orangutan Outreach, calling to him in a halo of light, he touched the screen, as if thinking, *He's talking to me.* Then, puzzled, he reached over and touched *Matt's* face. On the screen, a talking human, who knew him *by name*, was looking right at him and smiling, calling to him in a friendly voice. Why was Richard's face flat and Matt's face three-dimensional? He'd watched television lots of times, his favorite being nature films with orangutans. Matt sometimes showed him YouTube videos of adult male orangutans issuing their grown-up long calls, which always drew a fascinated stare. Yet the screen had never spoken *to* him. Hobnobbing with humans, rubbing shoulders with other orangs, meeting amiable strangers, playing with iPads, all had become staples of daily life. But this was an altogether different kind of socializing, and, although he didn't realize it, a step leading him deeper into the Human Age.

Parents today worry about the toll of screen time on their children's brains; the American Medical Association recommends none at all before the age of two. Yet a tech-enthused parent can even buy a child's "iPotty for iPad," a potty-training seat with built-in iPad holder, and find potty-training apps and interactive books at the iPad app store. Matt isn't concerned about Budi's iPad play, because unlike his own boys, Budi is a casual iPad user, and no one has studied the effect of screen time on the brains of orangutans. Would it make their senses and our own more alike? Anyway, because Budi is growing up surrounded by zoo life, human technology and culture will influence his brain in myriad ways, just as it does the brains of children. For good or bad, we use our rich imaginations to transfigure the world for ourselves and other creatures, banishing some critters we regard as "pests," while inviting others to share the curiosities we've invented (medicine, complex tools, food, special lingo, digital toys), urging them to blur the line with us between *natural* and *unnatural*.

IMAGINE, IF YOU like, Budi holding an iPad whose apps and games are chapters in this book. By touching the screen, he opens one, then another, merely listening to the story as human voices stream past, or watching colorful faces and vistas intently. In some he even catches a glimpse of himself, iPad in hand, as either an ape kid at play or a vital ambassador for his dwindling species. Both roles are his real-life destiny.

Lifting one hairy orange finger over the screen, Budi hesitates a moment, then touches the first chapter. When he does, a snowstorm opens up, with college students dashing between buildings, books clutched inside their parkas . . .

WILD HEART,
ANTHROPOCENE MIND

Knee-deep in the blizzard of 1978, when wind-whipped sails of snow tacked across Lake Cayuga, and the streets looked like a toboggan run, I was a student in upstate New York. Despite the weather, classes met, and scientists with souls luminous as watch dials were talking about nuclear winter, the likely changes in Earth's climate in the aftermath of a nuclear war: the sun white cotton in a perishable sky, dust clouds thickening over the Earth, plants forgetting how to green, summer beginning at twenty below zero, and then the seasons failing all living things. It seemed a possible scenario, since in Washington and Moscow, politicians were outdaring each other with playground bravado. This was the first time I'd heard my elders suggesting that we were now capable of unraveling the whole atmosphere shrouding Earth, and I was both wonder-struck and worried.

Only moments before, in geological time, we were speechless shadows on the savanna, foragers and hunters of small game. How had we become such a planetary threat? As the lectures and snow squalls ebbed, we students seemed small radiant forms in a vast white madness.

A quarter of a century later, Nobel laureate Paul Crutzen (who discovered the hole in the ozone layer and first introduced the idea of nuclear winter) stepped onto the world stage again, arguing that we've become such powerful agents of planetary change that we need to rename the geological age in which we live. Elite scientists from many nations agreed, and a distinguished panel at the Geological Society of London (the official arbiter of the geologic time scale) began weighing the evidence and working to update the name of our epoch from its rocky designation, Holocene ("Recent Whole"), to one that recognizes, for the first time, our unparalleled dominion over the whole planet, Anthropocene—the Human Age.

By international agreement, geologists divide Earth's environmental history into phases, based on ruling empires of rock, ocean, and life; it's similar to how we use "Elizabethan" and other royal dynasties to denote periods of human history. Deep ice cores in the Antarctic tell us of ancient atmospheres, fossil remains reveal ancient oceans and life forms, and more is written in silt and cataloged in stone. Previous periods, like the Jurassic, which we identify with dinosaurs, lasted millions of years, and we sometimes cleave them into smaller units, as changing epochs and eras slide into view. Each one adds a thread, however thin, to the tapestry. How wide a stripe will we leave in the fossil record?

PEOPLE WHO ARE recognizably human have walked the Earth for roughly two hundred thousand years. During those millennia, we survived by continuously adapting to our fickle environment. We braved harsh weathers and punishing landscapes, and feared animals much fiercer than we were, bowing to nature, whose spell overwhelmed us, whose magnificence humbled us, and around which we anxiously rigged our lives. After a passage of time too long to fully imagine, and too many impression-mad lives to tally, we began rebelling against the forces of nature. We grew handy, resourceful, flexible, clever, cooperative. We captured fire, chipped tools, hewed spears and needles, coined language and spent it everywhere we roamed. And then we began multiplying at breathtaking speed.

In the year 1000 BC, the entire world population was just 1 million. By AD 1000 it was 300 million. In 1500, it had grown to 500 million. Since then we've started reproducing exponentially. The world population has quadrupled since 1870. According to the BBC News website, when I was born, on October 7, 1948, I became the 2,490,398,416th person alive on Earth and the 75,528,527,432nd person to have lived since history began. In the Middle Ages, we were still able to count people in millions. Today there are 7 billion of us. As the biologist E. O. Wilson says, "The pattern of human population growth in the twentieth century was more bacterial than primate." According to Wilson, the human biomass is now a hundred times greater than that of any other large animal species that has ever existed on Earth. In our cities 3.2 billion people crowd together, and urban planners predict that by the year 2050 nearly two-thirds of the world's projected 10 billion people will be city-dwellers.

By the end of this decade, the history of planet Earth will be rewritten, textbooks will slip out of date, and teachers will need to unveil a bold, exciting, and possibly disturbing new reality. During our brief sojourn on Earth, thanks to exhilarating technologies, fossil fuel use, agriculture, and ballooning populations, the human race has become the single dominant force of change on the planet. For one species radically to alter the entire natural world is almost unprecedented in all of Earth's 4.5-billion-year history.

The only other time it happened was billions of years ago, before anything like golden-shouldered parakeets or marine iguanas, when the atmosphere was a poisonous brew and a time-traveling human would have needed to wear a gas mask. Then only spongy colonies of one-celled, blue-green algae blanketed the shallows, dining on water and sunlight, and pumping torrents of oxygen—their version of flatulence—into the atmosphere. Gradually, the air and ocean seethed with oxygen, the sky sweetened, and Earth welcomed creatures with lungs. It's a humbling thought, but one life form's excrement is another's tonic. For nearly five billion years, life ticked and tocked through an immensity of bold experiments, which the algae's recreation of the planet made possible, including all sorts of leaves and tongues, pedigrees and tribes, from Venus flytraps to humans. Then, improbably for origins so mundane, in roughly the last two or three hundred years, humans have become the second species to dramatically alter the natural world from earth to sky.

Humans have always been hopped-up, restless, busy bodies. During the past 11,700 years, a mere blink of time since the glaciers retreated at the end of the last ice age, we invented the pearls of Agriculture, Writing, and Science. We traveled in all directions, followed the long hands of rivers, crossed snow kingdoms, scaled dizzying clefts and gorges, trekked to remote islands and the poles, plunged to ocean depths haunted by fish lit like luminarias and jellies with golden eyes. Under a worship of stars, we trimmed fires and strung lanterns all across the darkness. We framed Oz-like cities, voyaged off our home planet, and golfed on the moon. We dreamt up a wizardry of industrial and medical marvels. We may not have shuffled the continents, but we've erased and redrawn their outlines with cities, agriculture, and climate change. We've blocked and rerouted rivers, depositing thick sediments of new land. We've leveled forests, scraped and paved the earth. We've subdued 75 percent of the land surface—preserving some pockets as "wilderness," denaturing vast tracts for our businesses and homes, and homogenizing a third of the world's ice-free land through farming. We've lopped off the tops of mountains to dig craters and quarries for mining. It's as if aliens appeared with megamallets and laser chisels and started resculpting every continent to better suit them. We've turned the landscape into another form of architecture; we've made the planet our sandbox.

When it comes to Earth's life forms we've been especially busy. We and our domestic animals now make up 90 percent of all the mammal biomass on Earth; in the year 1000, we and our animals were only 2 percent. As for wild species, we've redistributed plants and animals to different parts of the world, daring them to evolve new habits, revise their bodies, or go extinct. They've done all three. In the process, we're deciding what species will ultimately share the planet with us.

Even the clouds show our handiwork. Some are wind-smeared contrails left by globe-trotters in airplanes; others darken and spill as a result of factory grit loosed into the air. We've banded the crows, we've hybridized the trees, we've trussed the cliffs, we've dammed the rivers. We would supervise the sun if we could. We already harness its rays to power our whims, a feat the gods of ancient mythology would envy.

Like supreme beings, we now are present everywhere and in everything. We've colonized or left our fingerprints on every inch of the planet, from the ocean sediment to the exosphere, the outermost fringe of atmosphere where molecules escape into space, junk careens, and satellites orbit. Nearly all of the wonders we identify with modern life emerged in just the past two centuries, and over the past couple of decades, like a giant boulder racing ahead of a landslide, the human adventure has accelerated at an especially mind-bending pace.

Every day, we're more at the helm, navigating from outer space to the inner terraces of body and brain. We are not the same apes flaking tools on the savanna, toting gemlike embers, and stringing a few words together like precious shells. It's even hard to imagine our mental fantasia from that perspective. Did it feel more spacious or every bit as streamlike? We're revising the planet and its life forms so fast and indelibly that the natural world from which we sprang— atoms to single cells to mammals to *Homo sapiens* to dominance— is far from the same wellspring our ancestors knew. Today, instead of adapting to the natural world in which we live, we've created a human environment in which we've *embedded* the natural world.

Our relationship with nature has changed . . . radically, irreversibly, but by no means all for the bad. How we now relate to the land, oceans, animals, and our own bodies is being influenced in all sorts of unexpected ways by myriad advances in manufacturing, medicine, and technology. Many of nature's mysterious stuck doors have shivered open—human genome, stem cells, other Earth-like planets—widening our eyes. Along the way, our relationship with nature is evolving, rapidly but incrementally, and at times so subtly that we don't perceive the sonic booms, literally or metaphorically. As we're redefining our perception of the world surrounding us, and the world inside of us, we're revising our fundamental ideas about exactly what it means to be human, and also what we deem "natural." At every level, from wild animals to the microbes that homestead our flesh, from our evolving homes and cities to virtual zoos and webcams, humanity's unique bond with nature has taken a new direction.

I began writing this book because I was puzzled by certain questions, such as: Why does the world seem to be racing under our feet? Why is this the first year that Canada geese didn't migrate from many New England towns, and why have so many white storks stopped migrating in Europe? The world is being ravaged by record heat, drought, and floods—can we fix what we've done to the weather? What sort of stewards of the future planet will today's digital children be? What will it mean to travel when we can go anywhere on our computers, with little cost or effort? With all the medical changes to the human body—including carbon blade legs, bionic fingers, silicon retinas, computer screens worn over one eye with the ability to text by blinking, bionic suits that make it possible to lift colossal weights, and a wonderland of brain enhancers to improve focus, memory, or mood—will adolescents still be asking, "Who am I?" or "*What* am I?" How will cities, wild animals, and our own biology have changed in fifty years?

Without meaning to, we've also created much planetary chaos that threatens our well-being, the poor of the world's deserts and coastlines most of all. Yet despite the urgency of reining in climate change and devising safer ways to feed, fuel, and equitably govern our sprawling civilization, I'm enormously hopeful. Our new age, for all its sins, is laced with invention. We've tripled the average life span, reduced childhood mortality, and improved the quality of life for a great many people—from health to daily comforts—to a staggering degree. We are more globally aware now than ever. Our mistakes are legion, but our talent is immeasurable.

If we could travel back to, say, the Iron Age, few of us would go without packing certain essentials: matches, antibiotics, corrective lenses, compass, knife, shoes, vitamins, pencil and paper, toothbrush, fish hooks, metal pot, flashlight with solar batteries, and an array of other inventions that make life safer. We wouldn't travel light.

BLACK MARBLE

As our spaceship enters the roulette wheel of a new solar system, hope starts building its fragile crystals once again. Disappointment has dogged our travels, but we are nomads with restless minds, and this sun resembles our own middle-aged star. Like ours, it rules a tidy jumble of planets looping in atypical orbits, some unfurling a pageant of seasons, others hard-hearted, monotone, and remote. They're a strange assortment for siblings, with many small straggling hangers-on, but we've encountered odder night-fellows, and variety is their lure. One fizzy giant trails dozens of sycophantic moons; another floats inside a white cocoon. We weave between rocky, hard-boiled worlds, swing by a blimp tugging a retinue of jagged moons, dodge the diffuse rubble of asteroids, skirt a hothouse of acid clouds and phantom light.

Slowing to a hyperglide, we admire all the dappled colors, mammoth canyons of razor-backed rust, ice-spewing volcanoes, fountains fifty miles high, hydrocarbon lakes, scarlet welts and scourges, drooling oceans of frozen methane, light daggers, magma flows, sulfur rain, and many other intrigues of climate and geology. Yet there's no sign of living, breathing life forms anywhere. We are such a lonely species. Maybe this solar system will be the harbor where we find others like ourselves, curious, questing beings of unknown ardor or bloom. Life will have whittled them to fit their world, it doesn't matter how.

One more planet to survey, and then it's on to the next port of call.

On a small water planet flocked over by clouds, sequins sparkle everywhere. Racing toward it with abandon, we give in to its pull, and orbit in step with nightfall shadowing the world, transfixed by the embroidery of gold and white lights—from clusters and ribbons to willful circles and grids. *Crafted* lights, not natural auroras or lightning, but *designed*, and too many, too regular, too rare to ignore.

IN 2003, ABOARD the Space Station, Don Pettit felt his heart pinwheel whenever he viewed Earth's cities at night. *If only everyone could see Earth like this,* he thought, *they'd marvel at how far we've come, and they'd understand what we share.* A born tinkerer, he used spare parts he found in the Space Station to photograph the spinning planet with pristine clarity, as if it were sitting still. When he returned home he stitched the photographs together into a video montage, an orbital tour of Earth's cities at night, which he posted on YouTube. His voice-over identifies each glowing spiderweb as we sail toward it with him, as if we too were peering out of a Space Station window: "Zurich, Switzerland; Milan, Italy; Madrid, Spain.

"Cities at night are caught in a triangle," he says with awe tugging at his voice, "between culture, geography, and technology. . . . Cities in Europe display a characteristic network of roads that radiate outwards. . . . London, with a tour down the English coast to Bristol. Cairo, Egypt, with the Nile River seen as a dark shape running south to north, the Pyramids of Giza are well lit at night . . . Tel Aviv on the left, Jerusalem on the right . . ."

Glowing gold, green, and yellow, the Middle Eastern cities seem especially lustrous. He points out India's hallmark—village lights dotted over the countryside, softly glowing as through a veil. Then we fly above Manila, where geometrical lights define the waterfront. The dragon-shaped lights of Hong Kong flutter under us, and the southeast tip of South Korea. In the welling darkness of the Korea Strait, a band of dazzling white grains is a fleet of fishing boats shining high-intensity xenon lamps as lures.

"There's Tokyo, Brisbane, the San Francisco Bay, Houston," Pettit notes.

We don't intend our cities to be so beautiful from space. They're humanity's electric fingerprints on the planet, the chrome-yellow energy that flows through city veins. Dwarfed by the infinite dome of space with its majestic coliseum of stars, we've created our own constellations on the ground and named them after our triumphs, enterprises, myths, and leaders. Copenhagen ("Merchants Harbor"), Amsterdam ("A Dam on the Amstel River"), Ottawa ("Traders"), Bogotá ("Planted Fields"), Cotonou ("Mouth of the River of Death"), Canberra ("Meeting Place"), Fleissenberg ("Castle of Diligence"), Ouagadougou ("Where People Get Honor and Respect"), Athens (City of Athena, Greek goddess of wisdom). We play out our lives amid a festival of lights. The story the lights tell would be unmistakable to any space traveler: some bold life form has crisscrossed the planet with an exuberance of cities, favoring settlements along the coast and beside flowing water, and connecting them all with a labyrinth of brilliantly lit roads, so that even without a map the outlines of the continents loom and you can spot the meandering rivers.

The silent message of this spectacle is timely, strange, and wonderful. We've tattooed the planet with our doings. Our handiwork is visible everywhere, which NASA has captured with graphic poignancy in "Black Marble," its December 7, 2012, portrait of Earth ablaze at night. A companion to the famous "Blue Marble" photograph of Earth that appeared forty years ago, this radical new self-portrait promises to awaken and inspire us just as mightily.

On December 7, 1972, the crew of *Apollo 17*, the last manned lunar mission, shot the "Blue Marble" photograph of the whole Earth floating against the black velvet of space. Africa and Europe were eye-catching under swirling white clouds, but the predominant color was blue. This was the one picture from the Apollo missions that dramatically expanded our way of thinking. It showed us how small the planet is in the vast sprawl of space, how entwined and spontaneous its habitats are. Despite all the wars and hostilities, when viewed from space Earth had no national borders, no military zones, no visible fences. One could see how storm systems churning above the Amazon might affect the grain yield half a planet away in China. An Indian Ocean hurricane, swirling at the top of the photo, had pummeled India with whirlwinds and floods only two days before. Because it was nearly winter solstice, the white lantern of Antarctica glowed. The entire atmosphere of the planet—all the air we breathe, the sky we fly through, even the ozone layer—was visible as the thinnest rind.

Released during a time of growing environmental concern, it became an emblem of global consciousness, the most widely distributed photo in human history. It gave us an image to float in the lagoon of the mind's eye. It helped us embrace something too immense to focus on as a single intricately known and intricately unknown organism. Now we could see Earth in one eye-gulp, the way we gaze on a loved one. We could paste the image into our *Homo sapiens* family album. Here was a view of every friend, every relative and acquaintance, every path ever traveled, all together in one place. No wonder it adorned so many college dorm rooms. As the ultimate group portrait, it helped us understand our global kinship and cosmic address. It proclaimed our shared destiny.

NASA's new image of city lights, a panorama of the continents emblazoned with pulsating beacons, startles and transforms our gaze once again. Ours is the only planet in our solar system that glitters at night. Earth is 4.5 billion years old, and for eons the nighttime planet was dark. In a little over two hundred years we've wired up the world and turned on the lights, as if we signed the planet in luminous ink. In another forty years our scrawl won't look the same.

There are so many of us who find urban life magnetic that our cities no longer simply sprawl—they've begun to grow exponentially. Millions of us pack up, leave jobs and neighbors behind, and migrate to the city every year, joining nearly two-thirds of all the people on Earth. In the future, more and more clusters will appear, with even wider lattices and curtains of lights connecting them. Many display our curious tastes and habits. A harlequin thread drawn from Moscow to Vladivostok and dipping into China is the Trans-Siberian Railway. A golden streak through a profound darkness, the Nile River pours between the Aswan Dam and the Mediterranean Sea. A trellis connecting bright dots is the U.S. interstate highway system. The whole continent of Antarctica is still invisible at night. The vast deserts of Mongolia, Africa, Arabia, Australia, and the United States look almost as dark. So, too, teeming jungles in Africa and South America, the colossal arc of the Himalayas, and the rich northern forests of Canada and Russia. But shopping centers and seaports sizzle with light, as if they're frying electrons. The single brightest spot on the entire planet isn't Jerusalem or the Pyramids of Giza, though those do sparkle, but a more secular temple of neon, the Las Vegas Strip.

Newer settlements in the American West tend to be boxy, with streets that bolt north-south and east-west, before trickling into darkness at the fringes of town. In big cities like Tokyo, the crooked, meandering lines of the oldest neighborhoods glow mantis-green from mercury vapor streetlights, while the newer streets wrapped around them shine orange from modern sodium vapor lamps.

Our shimmering cities tell all (including us) that Earth's inhabitants are thinkers, builders and rearrangers who like to bunch together in hivelike settlements, and for some reason—bad night vision, primal fear, sheer vanity, to scare predators, or as a form of group adornment—we bedeck them all with garlands of light.

HANDMADE LANDSCAPES

Now let's zoom in closer.

The Earth isn't the same when you fly over it at three thousand feet and look for signs of humans. It's easy to lose your bearings. All the reassuring textures of daily life are lost. Gone are the sensuous details of wild strawberry jam, a vase of well-bred irises with stiff yellow combs, the smell of wild scallions beside the kitchen door. But it's a grand perch for viewing our tracks on the ground—visible everywhere and just as readable as the three-pronged Y's etched into the snow by ravens or the cleft hearts stamped by white-tailed deer.

The landscape looks very different than it did to our forebears, although we still use the sixteenth-century Dutch word (*lantscap*) to mean the natural scenery of our lives. Peering out of an airplane window, it's clear how we've gradually redefined that rustic idea. No longer does it apply only to such untouched wilderness as Alpine crags, sugared coastlines, or unruly fields of wildflowers. We manufacture new vistas and move so comfortably among them that quite often we confuse them with natural habitats. A field of giant sunflowers in Arizona or an extravagance of lavender in Provence offers a gorgeous naturalistic tapestry, even though both were sewn by human hands.

From the air, you can see how mountains lounge like sleeping alligators, and roads cut alongside or zigzag around them. Or slice clean through. Some roads curve to avoid, others to arrive, but many are straight and meet at right angles. Where forests blanket the earth, a shaved ribbon of brown scalp appears with implanted electrical towers shaped like stick men.

We not only bespangle the night, we broadloom the day. In summer, our agriculture rises as long alternating strips of crops, or quilted patchworks of green velour and brown corduroy. Miles of dark circles show where giant pivoting sprinkler systems are mining the water we unlocked deep below ground, which we're using to irrigate medallions of corn, wheat, alfalfa, or soybeans. Lighter circles linger as the pale shadows of already harvested crops. Evenly spaced rows of pink or white tufts tell of apple and cherry orchards. Among houses and between farms, small fragments of wooded land remain untouched: either the land is too wet, rocky, or hilly to build on, or the locals have set it aside on purpose to protect or use as a park. Either way, it proclaims our presence, just as the canals and clipped golf courses do.

Where retreating glaciers once dropped boulders and stones, scattering rocks of all sizes along the way, hedgerows border the crops. Farmers first had to unearth the rocks and boulders before they could till the land, and they piled the riprap along the edges of fields, where they were colonized by shrubs and trees that thrive in crevices and trap the drifting snow. On the first warm spring days, all of the snow will have melted from the corrugated brown fields, but not from the rocky white-tipped hedgerows that frame them.

Where dark veins streak the mountains, coal miners have clear-cut forests, shattered several peaks with explosives, scooped up the rubble, dumped it into a valley, and begun excavating. The blocks and crumbles of a stone quarry also stand out, and the terraced ziggurats of a copper mine rise above an emerald green pool.

Where mirages swim in the Mojave Desert's flan of caramel light, tens of thousands of mirrors shimmer to the horizon, each one a panel in an immense solar thermal facility. In other deserts around the world, and on every continent, including Antarctica, arrays of sun-catchers sparkle. Oil refineries trail for miles, swarmed over by pump jacks attacking the hard desert floor like metal woodpeckers and locusts.

Our pointy-nosed boats dot the ports and lakeshores; our tugboats wrangle commercial barges down the blue sinews of rivers. Newly hewn timber looks like rafts of corks floating toward the sawmills. Where marshlands attract flocks of migrating birds, one may also spot the scarlet paisley of our cranberry bogs, and the yellow of the mechanical growers that flood the bogs and then churn the cranberries to loosen them from their vines, corralling the floating fruit in long flexible arms. Red capital *T*'s are the stigmata of our evaporation ponds, where salt concentrates hard as it's harvested from seawater, in the process changing the algae and other microorganisms to vivid swirls of psychedelic hues. One sees our dams and harnessed rivers and the long zippers of our railway lines, and even occasional railway roundhouses. There's the azure blue of our municipal swimming pools, and the grids of towns where we live in thick masses piled one upon the other, with the tallest buildings in the center of a town, and long fingers of shorter buildings pointing away from them. The cooling stacks of our nuclear power plants stare up with the blank eyes of statues. Low false clouds pour from the smokestacks atop steel and iron plants, factories, and power stations.

These are but a few signs of our presence. Of course, our scat is visible, too. Junkyards and recycling centers edge all the towns, heaped with blocks of compressed metals and the black curls of old tires, swirling with scavenging gulls.

We've created a bounty of new landscapes, and lest the feat be lost on anyone, we even tack on the suffix "scape" to describe them. I've come across "cityscape," "townscape," "roadscape," "battlescape," "lawnscape," "prisonscape," "mallscape," "soundscape," "cyber-scape," "waterscape," "windowscape," "xeriscape," and many more. And let's not forget all the "industrial parks."

Although our handmade landscapes tend to fade into the background, just a stage set for our high-drama lives, they can be breathtaking. In Japan, tourists bored with volcanic mountains and gardens, and urban sightseers given to *kojo moe*, "factory infatuation," are flocking to sold-out tours that specialize in industrial landscapes and public works, which are viewed by bus or boat. Especially popular are the nighttime cruises that feature mammoth chemical factories spewing smoke and aglitter with star-clusters of light, overseen by the moon and more familiar constellations. It's become a popular date for romantic young couples.

"Most people are shocked to discover that factories can be such beautiful places," says Masakatsu Ozawa, an official in Kawasaki's tourism department. "We want tourists to have an experience for all the senses including that of factory smell."

"If you come to Tokyo, don't bother going to Harajuku," the city's shopping district, Ken Ohyama writes in his book *Kojo Moe*. "Go instead to Kawasaki," an industrial hub rich in rust, contaminated water, and polluted air. For that's where the industrial scenery is the most vivid. Some Japanese lawmakers would like a few of their working factories designated as World Heritage Sites, to draw even more tourists.

For the past twenty-five years, the Canadian photographer Edward Burtynsky has been documenting "manufactured landscapes" all over the world. Many of his most startling photographs were shot inside Chinese factories that ramble for blocks, where workers pass nearly all of their daylight hours surrounded by machines, products, and each other, under artificial light. The size and scale of their surroundings play upon the eyes and mind as a landscape. So does each floor of a large office building in, say, Singapore, divided into dozens of honeycomb cubicles.

I find Burtynsky's studio loft on a busy street in downtown Toronto. Large wooden tables flank several small offices, and a row of tall windows offers a portrait gallery of the day's weather. A tall, slender man with graying hair and neatly trimmed mustache and goatee greets me, and we retreat into his book-lined office. He's wearing a blue long-sleeved shirt with a small coyote logo howling up at his face. His voice is whisper-quiet, there's a calm about him almost geological in its repose, and yet his eyes are agile as a leopard's. ·

"You've been called a 'subliminal activist' . . ."

Burtynsky smiles. The moniker fits.

"Part of the advantage one has as a Canadian," he explains, "is that you're born into this country that's vast and thinly populated. I can go into the wilderness and not see anyone for days and experience a kind of space that hasn't changed for tens of thousands of years. Having that experience was necessary to my perception of how photography can look at the changes humanity has brought about in the landscape. My work does become a kind of lament. And also, I hope, a poetic narrative of the transfigured landscape and the industrial supply line. We can't have our cities, we can't have our cars, we can't have our jets without creating wastelands. For every act of creation there is an act of destruction. Take the skyscraper— there is an equivalent void in nature: quarries, mines."

Quarries as inverted architecture. I picture hollowed-out geometrical shapes, Cubist benches, ragged plummets. You can't have a skyscraper made out of marble or granite without a corresponding emptiness in nature. I haven't thought of our buildings in quite this way before, as perpetually shadowed by a parallel absence.

"And yet these 'acts of destruction' are surprisingly beautiful," I say.

"We have extracted from the land from the moment we stood on two feet. When we look at these wastelands, we say, 'Isn't that a terrible thing.' . . . But they can also be seen in a different way. These spots aren't dead, although we leave them for dead. Life does go on, and we should reengage with those places. They're very real and they're very much part of who we are."

My mind shimmies between two of his photographs: the stepped walls of an open-pit tungsten mine in northwestern Spain and a pyr-amid of lightbulb filaments, electronics, rocket engine nozzles, X-ray tubes, and the other particulate matter of our civilization. They're very different from the landscape photographs of the first half of the twentieth century, when Eliot Porter, Ansel Adams, and Edward Weston celebrated nature as the embodiment of the sublime, with reverence and respect, in all its wild untrampled glory. Burtynsky's photographs capture the wild trampled glory of humanity reveling in industry. For ages, nature was the only place we went to feel sur-rounded by forces larger than ourselves. Now our cities, buildings, and technologies are also playing that role.

Even calling something "nature" is a big change, Burtynsky sug-gests, from a time when nature existed all around and within us. Then we separated ourselves by naming it, just as, according to the Bible, Adam named the animals. Once we named them, they seemed ours to do with as we wished. Yet we were never as distant as we thought, and if we are learning anything in the Anthropocene, it is that we are not really separate at all. An important part of the landscape, our built environment is an expression of nature and can be more, or less, sustainable. The choice is ours.

IN THE HERE and now of an orangutan kid's life, Budi relinquishes his iPad for a moment. Then Matt lifts a hand, points down with his first finger, and swirls it around as if he were stirring up an invisible brew. On cue, Budi turns around and presses his back to the bars so that Matt can give him a scratch. Matt obliges, and Budi shrugs in pleasure, then presents one shoulder, arm, and back again for more.

"He just got his big-boy teeth a couple of months ago," Matt says. "His baby teeth fell out at the beginning of the year . . . he got rid of those giant Chiclets." Matt places some fresh fruit tidbits into Budi's mouth.

"He's very careful with your fingers."

"When he was really little he would bite—Hey, let go," Matt says, gently removing Budi's finger from a flap of iPad cover he's trying to pry off. "But then he had smaller teeth. When I'd squeal, he'd let go. Just like he was testing to see. He's a little bigger now, and even if he didn't mean to hurt me, he could."

They may be the same weight as humans, but orangutans are about seven times as strong, and may not realize the damage a playful yank or slap could do to a human. Yet they're also empathic enough to recognize another's pain, regardless of species, and feel bad about causing it.

"If he knows how to behave with people, the nicer his life's going to be—as he gets older he can do things like present body parts so that people can look after him. There's no guarantee that he'll be at this zoo forever, so it will be nice to say, *This is the language Budi knows. This is what you need to know to communicate with him.*"

Budi's mom, Puppe, wanders over to see what we're doing. Elderly by orangutan standards, at thirty-six, she's the oldest of the zoo's orangutans, with mature grayish skin (juvenile skin, like Budi's, is paler), a Buddha belly, and wrinkling around her nose and mouth. Her face looks strikingly humanlike, as does Budi's. Orangs meet our gaze with familiar faces and expressions across a hazy evolutionary mirage. Small wonder that, in Indonesian, their name means "Forest People."

Budi climbs the bars above his mom and dangles onto her head in a handstand, then slides upside down across her shoulders and rolls sideways off her back with a half twist. But she doesn't seem unduly bothered. After raising five tykes, she's used to such antics, and in any case she's always had a placid personality, a trait she's passed on to Budi, who tends to be relatively quiet as well. Not that orangs make much noise. The males may groan their long call to tell receptive females that they're hunks and other males not to mess with them, but the females and young always stay so close together that they only need to make subtle squeaks and grunts. Also, they're virtuosos of the visual. Most of their mutual knowing flows through an anatomy of signs, in which body language and pantomime offer a shared vocabulary. So Matt's work with them always includes gestures as well as words. It's a technique that's also gaining popularity among human parents with toddlers—teaching them basic sign language to make themselves understood before they can speak.

"Show me your tummy," Matt says, turning his attention to her and quietly gesturing *come here* with both hands.

"Let me see your tummy, Puppe," he says, pointing to her hairy orange belly. His tone with her is tender and respectful.

Puppe presses her big tummy close to Matt, who gives it a gentle rub. When he offers her some fruit she places a few pieces in one hand and delicately eats them one at a time.

"Where are you going, kiddo?" Matt says, as Budi runs off to a corner.

Grabbing a crinkly blue tarpaulin, he wraps himself up Caped Crusader style and returns to iPad play, triggering gorilla and rhino calls. Then Budi reaches for a control bar with buttons outside of the cage, and Matt brings the remote closer to him and lets him push the button that lifts a door on the wall dividing his enclosure from the next one. Hauling the tarpaulin overhead, he kicks a large ball through the door and dashes after it, brings it back, and pushes the button to close the door. Open, close, open, close. He's like any kid getting a rush out of opening and closing drawers and doors.

Matt believes in giving the orangs as much volition as possible, and lots of mental and sensory stimulation (or privacy if they wish).

"We make almost all their choices for them, and an intelligent animal should have opportunities to make more choices themselves," Matt says, "from deciding on the type of food they want that day to what activities they'd like to do."

"They didn't choose to be ambassadors for their ill-fated species," I think aloud, wondering if future geologists will discover that we allowed orangutans to go extinct in our age, or if we were able to rescue them at the eleventh hour.

"No." His face clouds over.

"The situation in the wild is very bad, I gather."

"The last I've heard," he says sadly, "is that the population is seg-mented, and right now none of the Sumatran orangutan popula-tions are sustainable in the long term, unless we can create corridors and protect those areas. There are so many benefits to orangutan corridors—they handle the storm water, they prevent erosion, they produce oxygen, they provide places for orangutans to live. The owners don't want orangutans near their palm plantations, but if there were functioning corridors, there would be less animal–human conflict."

So there's an Orangutan Awareness program at the Toronto Zoo, with education, outreach, and fund-raising for global orangutan projects. And there's the signature Apps for Apes program (at twelve zoos thus far) reminding people how much we have in common with the other great apes. When we see an orangutan at his iPad we naturally think, *He could be my son, my brother, myself.*

Budi touches a game on his iPad and the screen becomes an extravaganza of flurrying creatures, alive and finning, bubbling and whirling, in an underwater prehistoric world that Budi will never see. Nor will we, for we only know them at a standstill, as uninhab-ited bones, relics of a previous age as dramatic as our own.

A DIALECT OF STONE

'm wearing a fossil trilobite pendant around my neck right now. Black with prominent ribs in a silver bevel, it resembles a wood louse, and I wonder if it could rolypoly itself and somersault as wood lice do. Mostly, I wonder what its compound eyes saw so long ago. My trilobite is only an inch long, but I've held one nearly two feet wide in a neighbor's private collection, its ribs a xylophone impressive enough to play a tune upon. Trilobites are uncanny instruments of life.

Millennia before the Pliocene's celebration of the spine, when grazing quadrupeds roamed, silver birch leaves flickered like tiny salmon, and grebes first hinted at lunacy, trillions of trilobites prowled the ocean floors and paddled mud banks ajell with bacterial slime. In the evolutionary arms race, they grew armor plates, jointed legs, tough, chitinous jaws—anything to beat extinction's warrant. When they died, they bedded the muzzy swamplands. Today human bone-tumblers ogle their chalky remains, their exquisite herringbone shells. Trying to understand their habits, we sometimes allude to their cousins, the crab, spider, and millipede, and say "adaptive radiation about a common theme." As if that explained the papery organs within, or all the crises that fed their opportunity.

The most successful *water* animal ever embalmed as fossil, trilobites kept refining and upgrading themselves, over three hundred million years, until around twenty thousand different species freewheeled through deep and shallow seas on what must have seemed a trilobite-smitten planet. Some worked as stealthy ocean predators and scavengers, others as mild-mannered plankton-grazers, and still others fell in cahoots with sulfur-eating bacteria. Some developed protruding antlers and crackerjack spines. They scanned their realm with some of the oldest eyes on record, bug-eyed peepers with many lenses that weren't organic but mineral, made of six-sided crystal calcium prisms. These radically different eyes didn't provide crisp images but did offer a very wide field of view and motion. When a mass extinction wiped out trilobites 260 million years ago, their ancient lineage yielded to our world of insects with multifaceted eyes. But in their heyday, trilobites trolled the water world, and when they died their calcium carcasses fell to the bottom, crystal eyes and all, where layer upon layer of sediment enshrined them, gluing and compacting their bones with bits of coral and other calcium-cored creatures. Then time stacked its heavy volumes upon them, squeezing out the excess water and leaving behind limestone laced with skeletal remains. Today we use raw limestone in our roads, and grind it for paints and toothpastes—which means we use ancient trilobites, coral, and other fossils to help scrub our mouths.

That's also what it would take to fossilize humans—not as populous as trilobites but the most successful *land* animal ever. Just as well. I don't know how I'd feel brushing my teeth with the remains of ancient in-laws, or outlaws.

DRIVING DOWN THE highway that skirts Lake Cayuga, between glacial chunks of rock, I pass the uncanny work of erosion, a great sculptor of landscapes. Geologic eras are piled one on top of the other like Berber rugs, trilobites and other fossils bear witness to the evolution of life, and a host of creeks and waterfalls fume into the deep, gray-blue lake a thousand feet below. The wide ribbons of gunmetal gray and black shale I pass came from low-oxygen mud. Once this region was a shallow tropical sea that, as it evaporated, left not only mud full of marine-life skeletons that hardened into limestone but salt deposits, some of the deepest in the world. Colliding continents, 250 million years ago, stressed some of the rocks until fractures formed, land lifted, sea levels rose and fell. It's easy to forget, when you look out over the rolling hills, that you're seeing what once was the bottom of a sea, not the top of a mountain.

By the time dinosaurs appeared, 240 million years ago, the seas had retreated, leaving dry land, where dinosaurs stomped their footprints. When the ice age arrived, only 2 million years ago, it spread vast sheets of ice that repeatedly charged forward and dragged back, in the process gouging the deep Finger Lakes while streams cut the gorges of upstate New York. Sometimes large fossils appear from that era, like the mastodon (a hulking relative of the elephant with exceptionally long tusks) that a bemused local farmer found in his field several years ago.

Most of the life forms that once inhabited the planet have vanished, leaving no trace behind. Their remains have been polished down by the elements and bulldozed by the slow-motion avalanche of the glaciers. But geologists like Cornell University's Terry Jordan can read a tale in the rock strata, the Earth's dialect of stone, including the chevrons that tell of some sea-tossing event, maybe a hurricane like Sandy or Haiyan.

I like Terry Jordan from the first moment we meet in her office beside one of the sinuous plummeting gorges that are a hallmark of this lake district, a place for the geologically curious, loaded with fossils. It's her blue argyle socks—the crisscrossing design echoes the angles one sometimes sees in rock creased by spells of turmoil, the shadow of a buckling or churning calamity so long ago that we can only date it to within tens of thousands of years.

Specializing in sediment, the earth scruff that remains long enough to petrify into pinnacles and mutate into mesas, she's taught geology for much of a human lifetime. That seems long to me, but in her view of the past it's only a speck.

"Does it ever feel strange thinking in such long, slow units of time, when today's world is all about speed?" I ask.

Shaking her head with a laugh, she says, "No, I'm amazed by people who don't think this way. How can they not see it?"

The "it" is our place in the rocky bones of history. Admiring a chunk of rock on a table by Terry Jordan's window, I lift it in my hands and peer at the embossed fossils on its surface, where ram's-horn-shaped ammonites look like they're butting their way out.

"This is a wonderful place for fossils," I say. "Do you think our bones will show in the fossil record in this way, oh, say, ten million years from now?"

"Only if we're trapped in sediment!" she says, with a slightly impish smile. Then, seriously: "Maybe people living in coastal areas like New Orleans, Tokyo, or the Netherlands, or island nations—areas that will sink and disappear in mud when the sea level rises."

When we talk about the Age of the Dinosaur or the Age of the Trilobite, we expect to find fossils. But that's not true in the Age of Humans. It's not necessarily our bones that future geologists will ponder, but an altogether different kind of evidence. Not our bones but our residue will signal the beginning of the Anthropocene, a point delineated by a "golden spike"—a marker scientists pound into the rock strata to denote an internationally agreed-upon start of a geological time period. Most spikes are in Europe's heavily studied ribbons of exposed rock, with seven golden spikes in the United States, and dozens more throughout the world.

Let's suppose once more that we are astronauts, this time visiting Earth millions of years after humans have left to pioneer other worlds, allowing Earth to lie fallow for a spell and restore its bounty. Few signs of us remain—on the lush, overgrown surface, that is. Exposed rock and ice cores outline our story, and a future geologist—we'll call her Olivine—is looking for "time-rock," layers that show magnetic, chemical, climatic, or paleontological signs of the new age that we created.

From the warmth of her sky-tent tethered above what once was Patagonia, she travels the world and digs, measures, and tests, unearthing clues like shards of pottery. In sedimentary rock near the coasts, she pores over the fossil remains of cities: low-lying mazes once called Miami or Calcutta. She detects radioactive pulses from nuclear waste dumps. She finds a layer of woodland pollen suddenly replaced by agricultural pollen, and another ribbon where agriculture gave way to cities. Seams of concrete and metal abound. Scouting the oceans, she detects how, in our age, we plowed up the seabeds by bottom-trawling with large heavy nets dragged across the ocean floor, scooping up any marine life in their paths. Olivine is briefly envious. *They were the first generation of humans*, she thinks, *with the instruments and satellites to be able to measure how geology was changing during their own lifetime. What an exciting era that must have been.*

Everywhere she travels, she stumbles upon a mass of fossils of species far from their native habitats, and clumped together, on a scale unique in all of Earth's geological history. Long before we began shuffling life forms, species invaded new lands when continents collided in slow motion. But during our geologic moment, we've rushed the process, and quickly surpassed plate tectonics as a rearranger of species.

Olivine smiles as she identifies rose and Scotch broom pollen in a rock sample not far from her tent, in Patagonia, once a wild and windswept frontier where armadillos roamed and the beach pebbles were jasper. *Roses in Patagonia*, she thinks. *Those old Anthrops were rose addicts. And Scotch broom—didn't they realize it would spread for miles? Probably transplanted by settlers.* She knows the Scotch broom flailed long pokers of yellow flowers and, with weedlike momentum, colonized hundreds of acres and retuned the chemistry of the soil.

Special prizes are bits of human bones, whose DNA shows how our species began continent-jumping like hopping spiders, with some lineages predominating but most mixing, homogenizing, traveling far from their native shores, in a worldwide flux binding us all together.

She and her colleagues have argued some about the exact start of the Anthropocene—Agriculture? Industry? Nuclear bombs?—but they all agree that our world dramatically changed around the year 1800. That's when the Industrial Revolution, powered by a massive use of fossil fuels, led to rising carbon dioxide levels. We tend to forget that the steam engine was first invented to pump water out of coal mines, and only later adapted to move boats, cars, and trains. It's also when land clearing speeded up, and ecosystems were converted from mostly wild to mostly human-centered. Agriculture and mining became mechanized giants, spilling more fertilizer into the rivers and oceans, and more pollution into the air. The new textile mills and factory system drew laborers from the country into rapidly booming modern cities. That's when we first began adapting the planet to us on a large scale—changing the climate, changing the oceans, changing the evolution of plants and animals.

In the process, we've left our signature everywhere. Our impact is already measurable in the geologic record. The rocks hum with radioactive elements from atom bomb tests of the 1960s. The fossil pollen Olivine studies in the strata will reflect how, during our epoch, the wild brew of species that once thrived for centuries on the prairies and in the forests suddenly gave way to unbroken fields of single crops—corn, wheat, soy—and vast clans of cows, pigs, and chickens.

Because plastics take so long to degrade, they, too, will show up in the fossil record. Not as flattened lawn chairs and PVC pipes, but as veins of tiny plastic tears, which is as far as plastic denatures. Quasicrystals (crystals with orderly but nonrepeating patterns), transparent aluminum, and other newly invented forms of matter will appear in the matrix as well.

That alone is astonishing. We're adding new elements to the sum of creation. The wide world of nature, with all its chemicals and potions, plants and animals, rocks, crystals, and metals, is not enough for us.

We're also minting brand-new states of matter, metals no earthly eyes have ever seen—photonic clusters that can slice like light sabers, ultracold quantum gas known as polar molecules, fleecy electric, synthetic radioactive elements from einsteinium through copernicium to ununoctium, among many other artificial sprinklings. Monuments needn't be large, or even visible to the naked eye, to declare our godlike powers. It's one thing to rearrange bits and bobs of nature to create, say, an antibiotic or an atom bomb. But it's quite another to cook up exotic blazons of matter, adding them like new spices to the cosmic stew. I'm amused to think of a future geologist like Olivine puzzling over weird taffylike wads, trying to figure out what chem lab sport might have spawned them.

All of these elements will ultimately show up as indisputable signs of our presence. We're leaving tracks in the strata never seen before in Earth's five billion years. What did you add to the fossil record today? The plastic from a six-pack or a water bottle? A midden of candy wrappers, plastic bags, and orange juice cans? Did you drive your car? If so, you've changed the weather by a whisker, and that, too, will ultimately add to the patterns in the rock, a legacy of our meddling from deep sea to outer space.

MONKEYING
WITH THE WEATHER

How extraordinary that we've modified the whole big baggy atmosphere, where the carbon dioxide, now climbing to historic levels, is a third higher than even two hundred years ago. The synthetic fertilizers that plump up our crops churn out more nitrogen than all of the plants and microbes do naturally. Analyzing sedimentary core samples from Arctic lakes, future geologists like Olivine will see how we've addled the chemistry of the oceans and the air.

We're but one hotshot species on a planet squiggling with life, and yet we've grown powerful enough to befuddle the world's weather and sour all the oceans. That's the speed and scale of our influence. On land, humans figure as a geologic agent comparable to the relentless power of erosion or volcanic eruption, and in the oceans, our impact is on par with an asteroid's. The reef death we've caused will be visible in the fossil record. As a point of comparison, the last time reef death happened was sixty-five million years ago, when a real asteroid wiped out the dinosaurs and many other life forms.

Once you've glimpsed reef death, you don't forget its lunar landscape. I've always loved scuba diving and the cell-tickling feel of being underwater. Offshore in Jamaica, I once swam through a button collector's variety of vividly colored fish and was so spellbound that one hand automatically touched my chest and my eyes teared. My guide's eyes questioned me through the fishbowl of his face mask. There was no way to mime that I wasn't hurt or frightened, but jubilant, merely glad to the brink of tears. How do you scuba-sign wonder?

Are you in trouble? he signaled.

No, no, I answered emphatically. *I'm okay . . . My heart is stirred*—I put an open palm over my heart, then made a stirring motion in the water—*and my eyes . . .* I made a rain-falling movement beside one eye with my fingers.

Surface? he motioned, his knitted brow adding a question mark.

No! I signaled stiffly. *I'm okay. Wait. Wait.* I thought for a moment, then made the sign French chefs use in commercials, the gestural esperanto for *This dish is perfection,* making a purse of my fingers and exploding open the purse just after it touched my mouth. Then I swept a hand wide.

Even with the regulator stuffed in his mouth and his eyes distorted behind the faceplate, he made an exaggerated smile, yawning around the mouthpiece so that I could see he was smiling. He nodded his head in a magnified *Yes!,* then made an *Okay* sign with one hand and led me deeper, using his compass and surfacing once to check his direction by sighting the boat.

After a ten-minute swim, we suddenly came to a maze of underwater canyons thick with enormous sponges and coral fans, around which schools of circus-colored fish zigzagged. Plump purple sea pens with feathery quills stood in sand inkwells. Tiny tube worms—shaped like Christmas trees, feather dusters, maypole streamers, and parasols—jutted out of the coral heads. Sea relationships are sometimes like those in a Russian novel; a worm enters the larder of a fine, respectable coral to steal its food, and just stays there, unevicted. I moved my palm over a red-and-white-striped parasol, and in a flash it folded up its umbrella and dragged it back inside the coral. A game divers love to play with tube worms. Hocus-pocus and the tube worm vanishes.

On a coral butte just in front of us, a dark gorgonian jutted out between the canyon walls, its medusoid hair straggling in the current. I laughed. *That gorgonian's hair's like my own,* I thought. And then I remembered: *We're mainly saltwater, we carry the ocean inside us.* That was the simple, stupefying truth—as a woman, I was a minute ocean, in the dark tropic of whose womb eggs lay coded as roe, floating in the sea that wet-nursed us all. I pulled my mask up and washed my face with saltwater, fitted it back on, and exhaled through my nose to clear it. From then on, I was hooked, and often returned to the sea to reexperience the visible links of that invisible chain.

I was lucky. When I returned to that same spot twenty years later, I found the bare bones of a deserted reef, a moonscape.

There's no need to travel to the Caribbean to spot climate change's handiwork—I see it in my New York backyard. Perhaps you do, too, if you take the time to look closely. The looking closely part is essential. For most people, everything may still seem normal, because the seasons come and go in a familiar way, even if one blows in stormier or exits drier than usual. For many of us, the changes are too subtle to notice as we go about our lives.

But clues abound, and not just in my own backyard. Global warming is fiddling with garden thermostats to such an extent that the National Arbor Day Foundation has redrawn the U.S. Hardiness Zone Map—which tells gardeners what and when to plant. For thirty years (as long as the maps have been drawn), Ithaca lay in frostbitten, forget-about-lavender-hued-roses zone 5. Now most of New York State is in the warmer planting zone that used to lie farther south. The "what" and "when" to plant have changed, but not in a predictable way.

A row of ornamental cabbages (always annuals) has begun overwintering and sending up tall stalks of bushy yellow flowers for the first time. No one told the pansies, high summer blossoms, to call it quits in early winter. They keep blooming through snow showers, frost crackles, and quick melts . . . always with a pensive face. What became of all my Japanese beetles, those polychrome hedonists who used to mate in flesh piles, while eating, atop the roses? I haven't seen any for three years. But the number of Lyme ticks and other insects has soared. When I first moved to upstate New York decades ago, no Lyme ticks trickled through the grass; the cold climate was too hostile. They usually begin their blood-sucking on the white-footed mouse. Last summer's prolonged sizzle reduced the acorn crop, a mouse staple, and with fewer mice to hitch rides on and use as all-purpose canteen-nursery-gadabout-vectors for disease, the pesky parasite ticks began hopping aboard more humans. At least that's how it seemed; people venturing across a meadow inevitably returned with a Lyme tick in tow.

Imagine if you arrived home from work one day to discover that your pet spaniel had morphed into a wolf. You know that dogs evolved from wolves that we domesticated and hybridized . . . you just didn't expect to find one gnawing on the sofa leg. Something similar happened in my garden. A favorite yellow Canadian rose bush, well adapted to the cold climate, has been blooming faithfully and true for years. Like many other garden roses, it's a hybrid produced by grafting domestic and wild strains together. However, last summer, the rose suddenly revealed its lurking Id. To my amazement, from its feral heart it launched flutelike canes of heavily-flowering, tiny white roses. The wild rose ribs sprang from the same trunk as the well-bred yellow tea rose ribs. It was like having Siamese twins, one of which was Neanderthal, the other *Homo sapiens*.

Heaven knows what it will do this summer. Wild roses are hardier, better adapted to unstable temperatures. Will climate change favor one or the other? Will all of the domesticated roses run wild? A garden is always full of surprises. Last summer, for the first time in the decades I've lived here, my yard was a deafening amphibian rave, where hundreds of croaking frogs (especially the drum-eared bullfrogs, whose croak should really belong to a snoring bull, and the smaller banjo-plucking green frogs), bleatingly love-sick, drowned out human conversation. This year all I expect is the unexpected.

Canadian scientists warn of fewer backyard ice-skating rinks and frozen ponds in the future, and in some regions none at all, because of winter's waning bite. This inspired geographers at Wilfrid Laurier University in Waterloo, Ontario, to found a website to track the effect of a warm climate on Canada's tradition of thousands of icy flat playgrounds.

"We want outdoor rink lovers across North America and anywhere else in the world to tell us about their rinks," they urge on RinkWatch.org. "We want you to pin the location of your rink on our map, and then each winter record every day that it's skateable. We will gather up all the information from all the backyard rinks and use it to track the changes in our climate."

Many of Canada's legendary ice hockey players learned to skate on such tiny rinks, and Canadians hold them dear. An invisible thorn in the ozone layer can be denied, but when backyard hockey season is delayed, people notice.

Not everyone is warming up. Jim River, Alaska, a grizzly bear's backyard and a grizzled hiker's paradise, set a record low of -80°F. Residents there said the air hurt wickedly to breathe; they could feel it grate on every cell inside the nose. Exposed skin and eyes burned. Spit froze before it struck the ground. Frostnip took its toll. After a short spell outside, as people stepped back indoors, eyeglasses fogged up and froze to the face.

From Colorado to British Columbia, due to twenty years of unusually warm weather, spruce and pine bark beetles have chewed through four million acres of trees. This is fabulous for the bark beetles, but bad news for all the drought-weakened trees. Wildfires gust across their dry remains, sending flares through vast swaths of vegetation, as in the historic wildfires that blackened over 170,000 acres of caramel-mesa-ed New Mexico, and the record-breaking wildfires in mountain-blessed Colorado.

These massive conflagrations are bad not just for timber harvesters and tree lovers but for anyone who thrives on oxygen-rich air, since forests are the lungs of the planet, inhaling carbon dioxide and exhaling oxygen. We inhale their flammable waste to stoke the fires in our cells. They inhale ours. Bears, humans, and trees are as seamlessly connected as in and out breaths. And all this ash lies down quiet as snowfall, slowly settling to leave its trace, our trace, as the fire-debris weaves into the geological record. A fine line perhaps, but indelible as the cinders of Vesuvius.

Frostbite and torched forests may be the extremes, but 2012 and 2013 were legendary scorchers throughout the United States. Across the heartland, around the church suppers, cicada songs, and quiet nights of teenagers sitting on the paint-peeling white bandstands in the middle of town, frying heat doomed crops and broke 29,300 high-temperature records. Fall drought withered crops in 80 percent of the country's farmlands. Broad-brimmed-hatted, slow-drawling Texans saw the driest year since record-keeping began in 1895, drier even than the rawhide soil of the Dust Bowl. So dry that, as farms resorted to irrigation, public water supplies plummeted. The Lone Star State alone had $5 billion in damages. Not just from crop losses, either. The earth became so parched that it cracked all over like a callused heel, in the process wrenching apart water mains (forty in Fort Worth alone) and buckling the pavement on bridges and roads.

Worldwide, the past year ushered in record-breaking snowfalls, droughts, rains, floods, heat, hurricanes, wildfires, tornadoes, even plagues of locusts. The whole bag of tricks, biblical in their proportions, including weather pranks we usually expect, but not all and everywhere and wound up to such an extreme. Taken as a whole, as one weatherworks out of balance, it understandably starches the mind, widens the eyes, and fills parents with worry about their children's future. Every six years or so, the United Nations Panel on Climate Change issues a report. In September 2013, the panel of 209 lead authors and 600 contributing authors, from 39 nations, poring over 9,200 scientific publications, came to these landmark conclusions: global warming is "unequivocal," sea levels are rising, ice packs are melting, and if we continue at this pace we "will cause further warming and changes in all components of the climate." However, they added, we can slow the process down if we begin at once.

How the story plays out will be a tale told by the silent, everlasting rocks, in colorfully hued bandwidths. They'll recall a time when Earth was swarmed over by intelligent apes who whipped the weather into something they hadn't quite intended.

Yes, our tinkering has given Earth a low-grade fever, which we need to quickly calm before it climbs. But global warming won't be tragic everywhere and for every species. That would only be true if Earth's creatures, landforms, geology, waters, and climate were spread evenly around the planet, and they're not. Earth is a patchwork of many different habitats, and climate change will visit them in uncanny ways: cool hot zones, heat cool zones, flood dry zones, dry temperate zones. Thanks to climate change, Europe's growing season has been lengthening, with warm-season crops thriving farther north, to the delight of farmers (although in central and southern Europe, crops have suffered because of the extreme heat and drought). In Greenland, local farmers, seeing fertile soil for the first time, began avidly planting. Milder winters require less heating, which saves on energy, and travel and homesteading in the north is much easier in a warmer world. Not that long ago in the grand scheme of things, we had a famously balmy spell. During the Medieval Warm Period, from 950 to 1250, the Vikings found the lack of sea ice so good for travel that they established a colony in present-day Newfoundland.

A warmer world won't be terrible for everyone, and it's bound to inspire new technologies and good surprises, not just tragedy. Change is the byword everywhere, and if there's one unchanging fact about humans it's that we loathe change in nature, perhaps because we feel we can't control it. We may thrive on changes in technology and locale, but we want nature to be permanent and predictable, even when shaken, like the world inside a snow globe. We yearn for continuity, and yet we live in a wildly changing world. We love life fiercely, and yet we're creatures who die. These aren't reconcilable paradoxes.

We may not be noticing all of our leavings in the fossil record, but from the melting ice-skating rinks of Canada and the paling reefs of Samoa to dry creeks in Australia and receding glaciers in Chamonix, people are noticing the rude change in weather. We are beginning to see, firsthand, how our tinkering with the climate touches the globe from top to bottom. In my own extended backyard of New York State, the new normal recently wore the name of Sandy.

GAIA IN A TEMPER

The weather app Budi touches opens with a fright that his wild relatives have witnessed firsthand many times: torrential rains, snorting winds, and trees snapping—the familiar trees that orangutans mentally map for food and travel, just as we do houses, streets, and stores. Lately, though, whipped up by climate change, hurricanes like this one are growing to unforeseen and unimaginable fury.

A FREAK WINTER storm and tropical hurricane rolled into one, Sandy drew breath off Africa's west coast, barreled across the Caribbean, and charged up the eastern seaboard of the United States, swinging left with a gut punch that smashed in houses, sucked boats out of harbors and hurled them, masts and rigging flying, into front doors and garages.

Only a day before Halloween, the scene was beyond macabre, as if a Chagall painting had suddenly come to life in a 90 mph whirlwind of whizzing trees, animals, and objects. People unlucky enough to be caught outside were pulled sideways down the streets.

It was as if a monster were wrestling electrical lines to the ground, clawing up roads, turning neighborhoods into sandboxes. Piers and boardwalks crumpled like cardboard as the superstorm slapped them into the sea.

In this most densely settled area of the United States, prone to both hurricanes and nor'easters, record tides are usually measured in fractions of an inch. A major hell-raiser, Sandy even shattered the record for record-keeping—its tides had to be measured in feet. In one seaside community in Queens, after tidal surges beat the local record by three feet, a twenty-foot wave washed the whole research station into the ocean. In another town, the storm smashed furnaces and gas pipes, igniting fires that leapt from home to home, where doubly stunned residents found their first floors flooded and their roofs alight. The homes burned like surreal Fourth of July sparklers. In the beachfront town of Breezy Point, Queens, a blaze devoured 110 homes in one neighborhood while firefighters struggled to reach them through fast-flowing streets. All three regional airports shut down and canceled twenty thousand flights; Amtrak halted service to the whole Northeast Corridor. Forty-three million gallons of water gushed through the Brooklyn Battery Tunnel. The pounding ocean filled tunnels and subways and submerged lower Manhattan, where a flotilla of cars bobbed like colorful beetles.

I'll never get used to sweetheart names—Debby, Valerie, Helene—referring to land-scrubbing, wave-rearing, homewrecking, cyclonic mayhem. The name Sandy sounds like it belongs to an innocent, sun-kissed surfer. I'm not sure why we choose to domesticate cataclysmic violence in this way. It's too reminiscent of World War II pilots painting their girlfriends' names on warplanes, a paradox captured with lyric poignancy by the pilot and poet Randall Jarrell, who wrote, "In bombers named for girls, we burned / The cities we had learned about in school."

Before the frenzy was over, Sandy killed fifty people in the United States and sixty-nine in the Caribbean, flattened the homes and gutted the lives of thousands, and left millions more without food, water, or electricity. She also dropped three feet of wet snow in West Virginia and the Carolinas, and Tennessee received the heaviest snowfall on record. At times it seemed as if Gaia were so pissed off she finally decided to erase her workmanship, atomizing the whole shebang and flicking our Blue Marble back into the mouth of the supernovas where our metals were first forged.

Sandy is on my mind because it recently besieged my state, but 2012 also saw massive flooding in Australia, Brazil, and Rwanda; fifty major wildfires in Chile; wicked drought in the Sahel; record-setting cold, rain, and snow in Europe; and typhoons in China destroying sixty thousand homes. My head is still spinning from 2011's Tohoku earthquake and tsunami. Who can forget Louisiana's 2005 ordeal with Katrina? And, dwarfing all of these, Haiyan, the most powerful typhoon ever recorded, which charged through the Philippines in 2013 and killed over five thousand people.

New York and New Jersey had felt relatively safe, until Sandy rearranged their silhouettes, gouging inlets and bays, creating new marshes and sandbars, changing the map, literally and metaphorically. Climate change hits hard when it batters at childhood memories. Watching news footage of homes collapsing, over and over, I kept returning to the beacon of Atlantic City, where my family spent brief summer holidays. There were no casinos lining the boardwalk then, no fancy restaurants. But what a delicious, hot, sandy carnival it offered kids. The wide beach was duned deep with scorching sand that became soothingly damp a few inches down—the perfect consistency for sculpting.

The boardwalk held endless fascinations, including saltwater taffy vendors, Belgian waffles with whipped cream and strawberries, a penny arcade with a mechanical gypsy fortune-teller and Skee-Ball bowling machines, a kazoo-playing man in front of the 5 & 10, the giant Planters Peanut Man, the charcoal artists who did quick portraits, and the crablike processions of three-wheeled wicker chairs. Because the wooden planks were warped, riding over them became a bumbling, creaking amusement ride. Pushed along, we laughed as high-heeled women kept getting stuck in cracks between the boards. And then there was Steel Pier, with all of its amusements and its diving horse.

All the run-of-the-mills neighborhoods rely on, and the balm of meaning absorbed by homes, objects, streets, and piers—all gone.

Hurricane season brings a humbling reminder that, despite our best efforts and prophesies, nature remains unpredictable. Even aided by hindcasts, as forecasters call reading the entrails of past hurricane seasons to predict the future, we really don't know what stew of storms the Atlantic will dish up, especially now that we've dumped in strange seasonings. We can't yet predict the location of the next typhoon or tornado, even with all our high-tech weather instruments, any more than we know the final scores of the Caribbean's upcoming cricket matches.

For people living in coastal communities, the sea has always proved a generous or temperamental neighbor. But at least they knew broadly what mood swings to expect. Experts are duly unnerved. "Freak weather events happen, right, but twice in the last two years?" said meteorologist Jeff Masters. "I think something's up," he added. "I think we've crossed over to a new climate state where the new normal is intense weather events that kill lots of people."

YUP'IK ESKIMOS HAVE spent over a decade trying to relocate to higher ground. On the northwest coast of Alaska, only four hundred miles south of the Bering Strait, in the tiny Yup'ik village of Newtok, the residents can smell the salty breath of disaster, robed in liquid gray and pulling at their feet. Sabrina Warner suffers from a recurring nightmare: waking terrified to find the ice-clotted sea crashing in, washing the bed out from under her and collapsing her home. She and her young son swim for their lives, clinging to rooftops as their village is washed away. But there's nowhere safe to shelter. One roof slips from her fingers after another, until no harbor remains but the roof of the school, the largest building in the village, perched like a precarious osprey nest atop twenty-foot beams that have been driven into soft earth. And then that, too, is swallowed by the blue-black mouth of the sea.

This is a plausible nightmare. As the sun-reflecting ice melts, the planet is thawing much faster in the far north, where winters have warmed by 3°F since 1975 (double the world average). The widening riverbed and marshes of the Ninglick River, which snakes around three sides of the village, are tearing at its innards before pouring into the sea. Any day now the whole village and many neighboring indigenous communities will sink into the melting permafrost, as if it were white quicksand, to join the realm of polar bears and narwhals in the rich seams of Eskimo lore. By 2017, if the U.S. Army Corps of Engineers' predictions are right, numerous native villages along the northwestern coast and barrier islands will be in the same fix.

As America's first climate-change refugees, the Yup'ik have appealed to the state and federal government for help, but, according to international law, people only qualify as refugees if they're fleeing violence, war, or persecution. And federal disaster relief laws only grant money to repair infrastructure and damage *in place*, not to help with relocation after slow-motion disaster. Our humanitarian laws aren't keeping up with the Anthropocene's environmental realities. Robin Bronen, an Anchorage-based human rights lawyer and a frequent visitor to Newtok, is working tirelessly to change them.

"This is completely a human rights issue," she argues. "When you are talking about a people [the Yup'ik tribe] who have done the least to contribute to our climate crisis facing such dramatic consequences as a result of climate change, we have a moral and legal responsibility to respond and provide the funding needed so that these communities are not in danger."

When the residents of Newtok do move to their new town of Mertarvik on Nelson Island, a mere nine miles away, most will return to their familiar ice-coast life of subsistence fishing, but they'll be exemplars of a different epoch.

Tuvalu, a palm-frond island country in the South Pacific, has begun evacuating its people to New Zealand. And the president of Kiribati, an island nation of thirty-two atolls sprinkled across 3.5 million kilometers of ocean between Australia and Hawaii, is negotiating with Fiji to buy five thousand acres of land so that his population of over 102,000 can relocate. One Kiribati native, Ioane Teitiota, appealed for refugee status to Auckland, New Zealand, arguing that none of Kiribati's atolls is more than two meters above sea level, and therefore his life was endangered by the rising seas of global warming. The judge who heard his case found the claim novel, but not persuasive.

BRAINSTORMING
FROM EQUATOR TO ICE

S uch claims may soon become less novel, but fortunately humans are gifted with a taste for the novel that's leading us to some innovative responses—not only old-style mechanical solutions like fossil-fuel-powered Industrial Age steel floodgates, but promising new ways of addressing root causes and harnessing solar power.

Some countries have been stalwartly battling tides for years. Maeslantkering, one of the largest moving structures on Earth, is part of a network of sluices, dams, dykes, levees, and storm surge barriers protecting the Netherlands from the blustery North Sea. Galveston, Texas, is designing what will be the United States's longest sea wall, affectionately known as the "Ike Dike," to protect the low-lying city from impending floods. In Venice, the $5.5 billion MOSE Project—seventy-eight mobile, underwater steel gates—will isolate the Venetian lagoon from the Adriatic Sea and shield the city from floods.

In London, gated crusaders are already at their posts. Like a row of giant knights standing up to their necks in water, armor gleaming, weapons hidden, the Thames Barrier spans the river near Woolwich, downstream of central London. Behind each helmet looms an antique vision of the brain, all levers and hydraulics. Trapped in futuristic folded steel, the fixtures appear lighter than water, though in reality each gate weighs about 3,300 tons. Altogether they can close ranks beneath the waves to protect the city from storm surges.

Like so many other rivers, the Thames is not as wide as it used to be. We encrust the banks of our rivers with shops and houses, nosing out into the water, as if we long to be osmotically part of the current, in the process narrowing channels and putting our lives and property at risk. London has flooded famously in the past—307 people died in the 1953 flood—and with the world's sea levels rising, and the city steadily sinking at a foot a century, it needs not so much Knights Templar as Knights Temperature.

In mild days and years, the gates lie open, and you can sail between the five-story-high helmets, admiring their sleek beauty and wondering what London's son John Milton would have made of them in the 1600s. "They also serve who only stand and wait," he wrote in a sonnet about his blindness. He meant himself, serving God, but the Knights of the Order of the Barrier also stand and wait, reminding Londoners of climate change's reality, and how it will play out in their backyards, streets, and wharves if they don't take steps. Meanwhile gleaming guards defend the shores. Peer into their faces and darkness greets you. They will serve a few years more before their metal weakens and they're replaced by newer and abler paladins.

There is something ironic and fitting about using the very process that has led to trouble—the burning of fossil fuels—to forge the protectors to combat the ongoing sequelae of that very climate change, technological lords of deliverance we hope will protect us from ourselves. Better to harness the sun.

One of the most heartening solar stories comes from northwestern Bangladesh, home to the world's largest floodplain. Even though it's not raining at the moment, the humidity is nearly 100 percent, and the air feels thick as rubber. The monsoon season is ending, and though the sky can still rip with fierce downpours, children wearing colorful tunics and trousers hurry to the riverbank to board their solar-powered school boat. Older villagers wait for the health boat or library boat or agricultural extension boat. Thanks to one man's ingenuity and generosity of spirit, hope floats whenever it floods.

Mohammed Rezwan, an architect and climate-change activist who grew up here, ached to see his country being ravaged by ever-worsening floods. He's noticed that as temperatures rise and more snow melts in the Himalayas, more water surges across the Bangladeshi floodplain. Every year one-third of Bangladesh lies underwater, as if a giant eraser annually scrubs away the hand-drawn pictures of family life.

Rezwan decided that he didn't want to design buildings only to see them and whole communities washed away before his eyes. So, in 2008, he founded Shidhulai Swanirvar Sangstha (which means "self-reliance"), a nonprofit that deploys a fleet of one hundred boats with shallow drafts that can skim across the lowlands, serving as libraries, schools, health clinics, and three-tiered floating gardens. He persuaded local boat builders to outfit these traditional bamboo boats with solar panels, computers, video conferencing, and Internet access. The fleet also provides volunteer doctors, solar-powered hurricane lanterns, and bicycle-powered pumps. Solar batteries on each boat can power cell phones and computers, and people may recharge lamps to take home—provided their children keep attending school. Thus far, the project reaches 90,000 families, and it expects to reach another 81,500 families by 2015.

Because people become stranded and can't feed themselves in flood season, Rezwan invented a technique for them he calls "solar water farming." As he explains: "The system includes floating beds made of water hyacinth (to grow vegetables), a portable circular enclosure created by fishing net and bamboo strips (to raise fish), and a floating duck coop powered by solar lamps. It has a recycling system—duck manure is used as fish food, cold-water hyacinth beds are sold as organic fertilizer, and the sun energy lights up the duck coop to maintain the egg production."

So giggling children attend school, even during flood months, and their families can produce food and clean water despite the deluges. In this way, if monsoons or conflicts push people from their homes, the flotilla creates lifesavers of education, medicine, food, lighting, and communication.

Rezwan can't single-handedly fight climate change, but his brilliantly simple solution is helping people adapt. The words "adaptation" and "mitigation" are appearing more and more often in the lexicon of climate scientists, who use them to cover practical (and impractical) responses to climate change.

Remember blue-green algae, to whom we owe our oxygen-besotted lives? One controversial idea is seeding the Antarctic Ocean with iron to trigger the growth of such algae. Algae thrive by absorbing sunlight and carbon dioxide, which they use to forge chlorophyll, but for that they need iron. As algae soak up carbon dioxide from the atmosphere, they sink down to the ocean floor and die. Scientists aren't sure yet if widespread "iron fertilization" is safe for ocean animals, so they recently tried a small test, dumping iron powder into an Antarctic whirlpool (so that it wouldn't spread). Sure enough, a giant bloom of algae diatoms arose, sucking carbon dioxide from the air, and after a few weeks many diatoms carried pearls of CO_2 to the bottom and died. Would it work safely on a large scale? That's the big unknown. Geoengineering is a highly controversial plaything. We won't know except by trying, and a bad outcome could be deadly. We've already been geoengineering the planet for decades, unintentionally, by saturating the air with CO_2 and the oceans with fertilizer—not with good results.

Geoengineering and adaptation ideas run the gamut from shucks-why-didn't-I-think-of-that to plain nutty. The monochrome Earth method includes painting cities and roads white, covering the deserts in white plastic, and genetically engineering crops to be a paler color—all to reflect sunlight back into space. Or installing roof tiles that turn white in hot weather, black in cold. More bizarre tech fixes include firing trillions of tiny mirrors into space to form a hundred-thousand-mile sunshade for Earth, or building artificial mini-volcanoes that spew sulfur dioxide particles into the atmosphere to block sunlight. There's even been a "modest proposal" that we genetically engineer future humans to be tiny so that they'll need fewer resources.

At the other end of the globe, on the Norwegian coast, a colossal Carbon Capture Storage facility, owned jointly by Norway and three oil companies, is bagging carbon emissions before they're released into the atmosphere and storing them in underground vaults. Carbon prisons are still too expensive to be practical everywhere, and not worry-free, but many countries are following suit. In a quest for the first technology that can efficiently, *economically* pull CO_2 out of the air, Richard Branson is offering a tempting prize of $25 million.

Research by the cell biologist Len Ornstein shows that if the Australian Outback and the Sahara were forested, they'd absorb all of the CO_2 we're pumping into the atmosphere every year. Obviously not an easy venture, it's technologically possible. Grassroots indigenous, nongovernmental groups have already planted over fifty-one million trees in Africa. In time the forests, absorbing water from the soil and releasing it back into the atmosphere through their leaves, will generate their own clouds, rain, and shade, cooling things down, and providing the bonus of sustainably grown wood for their host countries.

Meanwhile, the eastern coast of the United States, from Boston to Florida, needs widespread sea barriers, preferably of sand, and also artificial barriers and gates wherever they're workable. They needn't be poetic knights of deliverance like London's. They might even be a version of the natural reefs and oyster beds that once flanked the American coast.

Trillions of oysters lined the eastern shores, building up knuckle-shelled beds that blunted the storm surges, capturing waves on the reefs, where they collapsed before they could blast down the estuaries. The Hudson River estuary was famed for the quality of its oysters. Oysters closest to shore also filtered the water, which made the habitat ideal for marsh grasses, whose root systems, clinging to the land, kept it from eroding.

We've largely destroyed that long, sweeping natural barricade. Now, as hurricane surges pummel cities and harbors, we're starting to realize what we've lost, not just in small innocent-seeming meals of wild-gathered shellfish but through toxic runoff from cities and farms. And it's the same the world over—85 percent of the world's oyster reefs have vanished since the end of the nineteenth century.

To protect New York City, the landscape architect Kate Orff favors an archipelago of artificial reefs built from piles of "rocks, shells, and fuzzy rope," to attract oysters, because oyster beds naturally act as wave attenuators. In time, the oyster-encrusted barriers would filter the water and also serve as a kind of ecological glue. "Infrastructure isn't separate from us, or it shouldn't be," Orff explains. "It's among us, it's next to us, embedded in our cities and public spaces."

BLUE REVOLUTION

"Mariculture," I say, floating the image of a vertical ocean garden in my mind, as I climb into a heavy, buoyant, safety-orange worksuit designed for extended periods on cold water.

"Think of it as 3D farming that uses the entire water column to grow a variety of species," Bren Smith says, closing his own suit over a black-and-red-checked flannel shirt and jeans, zipping the fish teeth of ankle zippers, and latching the belt. This is just the beginning of his vision for an elaborate network of small, family-owned, organic, and sustainable aquafarms arranged along the East Coast—oysters in beds under curtains of kelp—to help subdue storm surges while also providing food and energy to local communities.

Climate change is especially hard on fishermen and on farmers. The thirty-nine-year-old seaman sitting across from me in a dinghy on a frostbitten morning in Stony Creek, Connecticut, is both. Bren has a slender build with powerful arms and shoulders, a sign of his rope-heaving, cage-hauling trade. Although he now shaves both face and head, his plumage for years was natural red hair and long beard, hints of which remain. With his flame-orange watch cap, cinnamon five-o'clock shadow, and rusty-blond eyebrows, he is a study in reds, the long wavelengths of visible light.

We're not anticipating a stumble overboard, but like many a fisherman Bren doesn't swim, and the suit adds needed warmth through high winds and snow-thunder in the recent cannonade of winter storms.

A perennial mariner, he grew up in Petty Harbour, a five-hundred-year-old Newfoundland town with eleven painted wooden houses filled with fisherfolk and a salt-peeled wharf with jostling boats. On the rocky shore, a boy could find lobster cages, floats, anchors, ropes, seaweed-tangled shells, fish and bird skeletons, and tall tales. So it's not surprising that, at fifteen, he dropped out of high school and ran away to sea. In Maine he worked on lobster boats, in Massachusetts on cod boats, and in Alaska's Bering Strait on trawlers, longliners, and crab boats. At one point he factory-fished for McDonald's.

"Do you think of yourself as a fisherman or a farmer?" I ask.

"A farmer now. It's more like growing arugula than facing the dangers of the sea—which, believe me, I've seen."

In a sense 3D farming is rotational agriculture. Bren harvests kelp in the winter and early spring; red seaweed in June and September; oysters, scallops, and clams year-round; mussels in the spring and fall. At least that's the theory. Hurricane Irene tore up his oyster beds, which he promptly reseeded, knowing he'd have to wait two more years for harvest. Hurricane Sandy smothered the oyster beds yet again. Clams have a better chance of surviving a hurricane because they at least have a strong foot and can move a little. But oysters really are trapped. They don't even move to eat or mate. Without the reefs, storm surges churn them up, and as the silt smothers the oysters they die, beginning the slow process of joining the fossil record. Right along with the Model Ts that sank when the Long Island Sound froze over in 1917–1918 and foolhardy souls tried driving across it.

"Ironically," Bren says thoughtfully, "I may be one of the first green fishermen to be wiped out by climate change."

But Bren is upbeat and confident. Fortunately, he was able to harvest some mussels in the thick of a snowstorm, just before Blizzard Nemo hit. Kelp, at least, is a post-hurricane-season crop. After Sandy he began planting the year's kelp, and now, in mid-February, it's nearly ready for harvest.

Unmooring the dinghy, Bren hops back in, and we motor out to his solar-powered fishing boat, placid as a tiny icebreaker half a mile offshore. En route, we weave through the Thimble Islands, an archipelago of islets, some with majestic cliffs of 600-million-year-old pink granite. Many are topped by stilted, turreted, luxe storybook houses with long wooden staircases winding down to the water.

A receding glacier left behind this spill of islands: massive granite knobs, stepping-stone slabs, and submarine boulders and ledges, some of which only appear at low tide. Named after wild thimbleberries, not thimble-sized cuteness, the cluster includes Money Island, Little Pumpkin, Cut-in-Two Island, Mother-in-Law Island, Hen Island, and East Stooping Bush Island, among many others— between 100 and 365 (depending on the height of the tide, how you define an island, and if you cherish the idea of an island for each day of the year), with around twenty-three of them inhabited by people during the summer. Harbor seals and birds abound. Each island is cloaked in its own gossip and lore, thanks in part to famous sojourners, from President Taft to Captain Kidd and Ringling Brothers' Tom Thumb.

As saltwater and river water mix in the estuary, it offers a feeding and breeding refuge to 170 species of fish, 1,200 species of invertebrates, and flocks of migratory birds. Horse and Outer islands are wildlife preserves. For Bren it's a fertile garden visited in summer by flocks of seasonal guests and in winter by tumultuous storms, but always spawning life above and below the surface. Today, in arc-light winter, with a chill wind slicing around the water streets, the garden is icy-blue and glaring, with air that's clear as a bugle call.

"The granite cliffs are amazing," I say, inhaling their feline beauty. Flecked with velvet-black biotite and streaks of cream and gray quartz, in the speckled sunlight, with the boisterous sea slapping at their base, they look more animal than mineral.

"It's the same pink granite that helped build the Statue of Liberty," Bren explains, "and the Lincoln Memorial and the Library of Congress. In Ayn Rand's *The Fountainhead*, the architect stands at the edge of a local granite quarry . . ."

"Full of capitalist machismo, as I recall."

"Exactly!" Bren says, blue eyes flashing. "I came here in part to erase that image and that extreme ideology."

I know the passage he means, the one in which Howard Roark, clothed only in his grandiosity, stands above the quarry, with all of nature his raw material, something to be devoured by the few powerful men who deserve to rule the world:

> These rocks, he thought, are here for me; waiting for the drill,
> the dynamite and my voice; waiting to be split, ripped, pounded,
> reborn; waiting for the shape my hands will give them.

"I pillaged the seas," Bren admits in a conscience-stricken voice. "When I look back over my life, I see it as a story of ecological redemption. I was a kid working thirty-hour shifts, fishing around the clock, and I absolutely loved it because I got to be on the open sea. But, you know, we scoured the ocean floor, ripping up whole ecosystems. We fished illegally in protected waters. I've personally thrown tens of thousands of dead bycatch back into the sea. It was the worst kind of industrial fishing."

There was a time when cod grew large enough to swallow a child. But fishermen have been systematically harvesting the largest fish, and the cod had to mate earlier and at a smaller size to survive as a species. The successful ones passed on their genes. Now a cod will fit on a dinner plate. Soon there will only be small fish in the sea. In the process of reducing them we've also remodeled our vision of cod—from a behemoth that could feed a whole family to a small and harmless fish. Even those are vanishing, along with other marine life forms, in one of the greatest mass extinctions ever to befall the planet. For Bren, the whole foraging, hunter-gatherer mentality has led to decades of what he thinks of as a kind of piracy, minus the romance.

"I went back to Newfoundland once it was clear to me that fishing like that wasn't sustainable. I loved the sea and I could see the destruction, and I became much more conscious of the ecosystem. After that I went to work on some of the salmon farms, but I saw the same sort of industrial farming. Not good for the environment, and not good for people. Wild fishing and farming fish—neither one was sustainable. The sea was in my soul; I knew I needed to work on the sea. But I was part of a new generation that wanted something different. So how could I evolve into a *green* fisherman, I wondered?

"I ended up here in Long Island Sound right at the time there was a movement to bring young fishermen under forty back into the fisheries. They opened up shellfish grounds. You see, it's very hard to get shellfish grounds because they're all owned by about six families going back generations. But when they opened up these grounds ten years ago, I came and started aquafarming. I thought, okay, on this sixty-acre plot of ocean, what species can I choose that will do several things to create sustainable food in a good way? And can I think beyond that and actually restore the ocean while we're farming it, and leave the world better than we started, but also grow great food?

"Suddenly I found myself growing food in the most efficient, environmentally sustainable way possible—vertically. And it grows quickly. The kelp will grow eight to twelve feet in a five-month period. And the whole food column is nourishing. The oysters, mussels, and scallops provide low-fat protein and all sorts of important vitamins: selenium, zinc, magnesium, iron, B vitamins, omega-3s. We've analyzed the sea vegetables—different forms of algae like kelp—and they create lots of vitamins and minerals and nine different amino acids, plus omega-3s. Could you actually have something called 'ocean vegetarianism'? I think so. During World War II, both the Germans and the British came up with this plan to deal with starvation, which they thought was going to be a huge risk in World War II, and actually they did all these studies and began feeding people algae. There's also some modern research that if you created a network of small seaweed farms around the world that added up to the size of Washington State, you could feed the whole world. Now, you're not going to get everyone to eat seaweed, but it shows the potential that's there.

"This is *Mookie*," he says, as we pull up to his cobalt-blue fishing boat with a sky-blue cabin door and a white deck sole that must once have matched. Thanks to the rubbing of boots, cages, ropes, and splintery dock, a trifle of paint has worn away to reveal a thin deckle-edge of sky blue.

We climb aboard, hoist anchor, and chug to his patch of ocean, a flowing field dark as gravy. Small gobs of sea spit trail the boat. Gray-and-white herring gulls spiral above, following us as they would any large predator, their yellow eyes hunting for small fish churned up to the surface.

Dropping anchor, we winch up a heavy cage and swing it carefully onto a built-in wooden bench. The cold breeze, snorting and blowing, is full of turning knives. I'm glad of the heavy worksuit, but it's cumbersome and my movements feel moonwalk slow.

Bren pops the lid to reveal a vault of about three hundred oysters and a mix of sea creatures, including starfish, small fronds of orange algae, and a necklace of round off-white periwinkle egg capsules that look like buttons of horn or coral.

"Look at this," he says, slicing an egg open with his teeth and extracting tiny seeds on the tip of a knife blade. "They're snail eggs, and they actually look like miniature snails."

Amazingly, they do. Periwinkles, flavorful sea snails, have been part of English, Irish, Asian, and African cuisine for millennia. Clinging to rocks (or oyster cages) to steady themselves, they feed on phytoplankton. But these freeloaders aren't welcome among the oysters. Nor is the squishy round sea squirt, or the translucent segmented mantis shrimp, or the cascade of olive-green sea grapes, or the broken shells. Back they all return to the sea, except for the mass of tiny open-jawed barnacles encrusting the mesh cages. Those have mortared themselves in place and will have to wait to sink with the oysters.

Gulls swim through the sky as we pour the oysters into shallow bins on a wooden table. Our job today is to "rough up" the oysters— not injure, but stress them so that they'll form tougher shells. Much as muscles build if you exercise them, oysters thicken their shells when tossed by the tide. Idle oysters need exercise, just as idle humans do. Without struggle, strength won't grow. The human parallel plays with my mind; then the cold blows the thought away, and I reach for a pair of rubber gloves.

"Knead them like bread," Bren says, showing me how.

Catching a dozen or so in my open fingers, I roll them forward with the base of my palms, then claw them back gently and repeat the undulating motion. I have become the tide.

"We touch them every five weeks," he explains, "to make sure they'll grow strong."

"It sounds like you feel pretty close to them."

"They're like family. I plant them, I'm with them for two years, watch them grow, touch them regularly. I know every oyster personally."

"By name?"

He laughs. "Not quite. Not *yet*."

"They're really beautiful." I pause to pick up one of the Thimble Island Salts and look at its deeply cupped shell and golden hue, purple patches, iridescent luster. Some resemble a bony hand, others a craggy mountain range.

When Bren opens one with a knife and offers it to me, I can't refuse. Oyster-proud, he waits for my response. Not an oyster con- noisseur, I just let my taste buds speak: "Incredibly salty, silky, smooth, plump as a mitten. It tastes like a bite of ocean."

A Proustian memory transports me to the coast of Brittany, in the shadow of Mont Saint-Michel, where people also harvest the sea.

There are huge tides there, and the water is very saline—perfect conditions for raising oysters, one of the great delicacies of Brittany. I remember them tasting salty, too, but different, slightly metallic with a whisper of tea and brass. Michel de Montaigne thought oysters tasted like violets. But the flavor of oysters varies depending on their environment, and I've read of some that leave an aftertaste of cucumber or melon.

"Good." He smiles. "If one doesn't taste good I feel like a failed father."

Returning the roughed-up oysters to their cage, we lower them back into place, swing the boat around, and check on the kelp dangling from black buoys along a hundred-foot line.

"Walking the line," Bren says, as he eyes each string of kelp prayer flags, barely visible beneath the cloud-shadowed water. Snagging one up with a red-handled hook, he hoists it out of the water, and I'm surprised to see a long array of curly-edged kelp ribbons, about three inches wide and a yard long, some with faint moiré stripes. Like land plants, kelp photosynthesizes, but not just the leaves, the whole kelp. As a result, it pulls five times more CO_2 from the air than land plants do.

A strand feels surprisingly dry and smooth, and sunlight glows through its golden-brown cheek. A longtime staple in Asian cultures, kelp (and other algae) adds depth to Canadian, British, and Caribbean cuisines. It's also been harvested for medicinal use since ancient times. Suffused with minerals, more than any other food, it harbors most of those found in human blood and benefits thyroid, hormone, and brain health. It also boasts anticancer, anticoagulant, and antiviral properties. It's the "secret ingredient" in the posh La Mer line of skin creams, among others. Its alginates are used to thicken everything from pudding and ice cream to toothpaste, even the living cells poured by 3D bioprinters.

"Try some," he says, offering me course two.

I taste a piece of kelp curl, which is chewy and rather tasteless, more texture than flavor, but perfect for noodling with sesame oil or in miso soup, as I've often eaten it in Japanese restaurants. Bren sells oysters and kelp to local residents and restaurants and to chefs in Manhattan.

"I think of this actually as 'climate farming,'" Bren says, "because the kelp soaks up huge amounts of carbon and can easily be turned into biofuel or organic fertilizer. So I'm in conversation with companies, NGOs, and researchers right now. Kelp is over 50 percent sugar. The Department of Energy did a study that showed if you took an area half the size of Maine and just grew kelp, you could produce enough biofuel to replace oil in the U.S. That's stunning! And without the negatives of growing land-based biofuel, which by the way is actually terrible. It wastes a lot of water, fertilizer, and energy. But here you can have a closed-energy farm, using zero fresh water, zero fertilizer, and zero air, while providing fuel for local communities. I grow this kelp here for food, but you could plant it in the Bronx River or in front of sewage treatment plants, which would reduce their polluting. Or you could grow kelp for biofuels.

"Over the past ten years I've been struggling with all of these things and trying to figure out how they could come together. Think about it. Growing food in the ocean: no fertilizer, no air, no soil, no water. None of these things that are hugely energy-intensive and huge climate risks to both freshwater and soil. When you put all of this together it's so exciting. It's so exciting! I can almost smell the possibility of a blue revolution joining the green revolution. And because it's vertical farming, it will have a very small footprint."

Not everyone agrees with his methods, especially old-style environmentalists, which he's the first to point out.

"Now there's a real pushback, of course, from some conservationists, because people think of the oceans as these beautiful wild spaces—which I'm so sympathetic to because I've spent my life on the ocean. But we're facing a brutal new reality," he says, his face aflame with resolve. "If we ignore the greatest environmental crisis of our generation, our wild oceans will be dead oceans. Ironically, climate change may force us to develop our seas in order to save them. We need to do that *and also* reserve large swaths of the oceans as marine conservation parks. This won't solve every problem we're facing, but it will begin to help."

Behind all of Bren's enthusiasm is a wave of widely shared concern about how climate change is acidifying the seas. He's part of a transitional generation that feels the urgency of reconciling their lifestyle with the planet's health. Call it what you will, pioneering or bioneering, because of his commitment, he was invited to join the Young Climate Leaders Network, which supports a small group of "innovative leaders and visionaries, including many who operate largely outside of the traditional environmental community, working for climate solutions."

Bren's eyes rest on the water. "There's no doubt, this will mean reimagining the oceans, which is heart-wrenching and controversial for a lot of people who revere the oceans as some of the last wild places on Earth, places untouched by human hands."

Yet the truth is that oceans are not untouched by human hands. In 2007, owners of the only salmon farm in Ireland woke one day to find its hundred thousand salmon devoured by a horde of jellyfish. Throughout the world's oceans, trillions of umbrella, parachute, and bell-shaped jellyfish have been swarming, lured by rising temperatures, nutrient-rich agricultural runoff, and pollution. With semitransparent stealth, they sneak up on flounder, salmon, and other large fish favored by human fishermen and colonize a slew of habitats, where they eat or oust the local fish. Oceana Europe, which works to restore and protect the world's oceans, attributes the soaring number of jellyfish to climate change and the human overfishing of tuna, swordfish, and other natural predators. City-dwellers are combating blooms of jellyfish in Tokyo, Sydney, Miami, and other harbors. During one recent summer, record numbers invaded the shallows of South Florida and the Gulf of Mexico. In Georgia, on one Saturday alone, Tybee Island Ocean Rescue reported two thousand serious stings.

The sea is a spirit level, a pantry, a playground, a mansion rowdy with life, a majestic reminder of our origins, another kind of body (a body of water), and female because of her monthly tides. But her bones are growing brittle, her brine turning ever more acidic from all the CO_2 we've slathered into the air and all the fertilizer runoff from our fields. While that's terrible for creatures like coral, oysters, mussels, and clams, whose calcium shells can soften and dissolve, the warmth is a tonic for starfish, which are roaming farther north in throngs. Until, that is, their shellfish prey vanish.

"Environmentalists have been asking the wrong question," Bren says after a moment. "It's not just about: How can we save the oceans? How can we protect the sea animals? I agree, all of that's important. But we also need to flip our way of thinking and ask: How can the oceans save *us*? How can it provide food, jobs, safety, and a sustainable way of life? I'm convinced the answer is ocean conservation with symbiotic green farms."

Last thing, we check the remaining crop of mussels, which means back-straining, heave-hauling them up from the depths where they're filling their mesh socks nicely, growing through the lattices like shiny black buttons, still too small for harvesting. So back they descend, too young for saffron cream sauce. I can see why he finds this part of his workday like checking on a nursery.

Scanning the lapping ripples of the Sound, it doesn't look like an industrial landscape at all. And yet the amount of food growing below the water is incredible. There are two tons of kelp on Bren's longlines alone. I like Bren's "symbiotic" way of thinking. We billions of creative, problem-solving humans don't have to be parasites in our environment—we have the technology, the understanding, and the desire to become ecologically sustaining symbionts.

On our return to Stony Creek harbor, we again pass the island-perched village of Victorian mansions and salt-white cottages, with stone chimneys for burning up yesterday's disappointments, rain-rattled windows, sea-spying porches, and wind-worn trees and gardens. And always the deep and dazzling blue of the Sound, with hidden reefs and ledges, devious currents corkscrewing just below the surface, and, during storms, waves running like greyhounds.

The new dock looks trim, clean, and stubbornly well anchored against hurricanes. A pair of black cormorants perches on a rocky knob, and Bren gestures a welcome. Superstition tells of drowned fishermen returning as hungry cormorants, dressed in black rain gear, with webbed feet instead of boots.

Despite the cold breeze there's a warm afternoon sun. Soon the tide will be walking in and the pink-legged seagulls skimming the shoreline. In a few months the summer crowds will arrive to eat fresh seafood, attend the puppet theater, fall asleep to the slurred voice of the ocean, and enjoy the ecstasy of coastal life and clean water, with time strapped to their wrists.

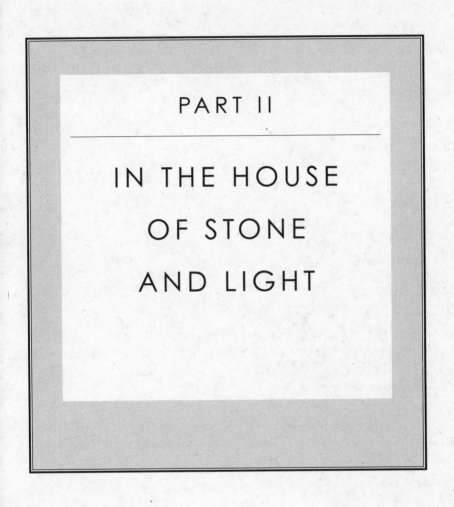

PART II

IN THE HOUSE
OF STONE
AND LIGHT

ASPHALT JUNGLES

Watching Budi tumbling and climbing, at play with ball and shadow and iPad alike, I marvel at the road the human race has traveled. Open your imagination to how we began—as semiupright apes who spent some of their time in trees; next as ragtag bands of nomadic hunter-gatherers; then as purposeful custodians of favorite grains, chosen with mind-bending slowness, over thousands of years; and in time as intrepid farmers and clearers of forests with fixed roofs over our heads and a more reliable food supply; afterward as builders of villages and towns dwarfed by furrowed, well-tilled farmlands; then as makers, fed by such inventions as the steam engine (a lavish power source unlike horses, oxen, or water power, and not subject to health or weather, not limited by location); later as industry's operators, drudges and tycoons who moved closer to the factories that arose in honeycombed cities beside endless fields of staple crops (like corn, wheat, and rice) and giant herds of key species (mainly cows, sheep, or pigs); and finally as builders of big buzzing metropolises, ringed by suburbs on whose fringes lay shrinking farms and forests; and then, as if magnetized by a fierce urge to coalesce, fleeing en masse into those

mountainous hope-scented cities. There, like splattered balls of mercury whose droplets have begun flowing back together, we're finally
merging into a handful of colossal, metal-clad spheres of civilization.

Among the many shocks and wonders of the Anthropocene, this
is bound to rank high: the largest mass migration the planet has ever
seen. In only the past hundred years, we've become an urban species. Today, more than half of humanity, 3.5 billion people, cluster
in cities, and scientists predict that by 2050 our cities will enthrall 70
percent of the world's citizens. The trend is undeniable as the moon,
unstoppable as an avalanche.

Between 2005 and 2013, China's urban population skyrocketed
from 13 percent to 40 percent, with most people moving from very
rural locales to huddled megacities whose streets jingle with chance
and temptation. At that pace, by 2030, over half of China's citizens
will live in cities, and instead of farming food locally they'll import
much of it from other nations, paying with the fruits of industry,
invention, and manufacturing. That's already the case in the U.K.,
where by 1950 a checkerboard of cities embraced 79 percent of the
population. By 2030, when the U.K.'s city-dwellers reach 92 percent,
it will be a truly urban nation, joining a zodiac of others. Ninety
percent of Argentinians already dwell in cities, 88 percent of Germans, 78 percent of the French, 80 percent of South Koreans. For a
rural nation, one needs to journey to Bhutan, Uganda, or Papua New
Guinea, where nearly everyone lives in the countryside, with a scant
10 percent committed to metropolitan life—so far.

Our city-chase has reached such a frenzy that the idea of *migration*
doesn't begin to capture its rush or rarity. This isn't surprising in an
economically lopsided world where, too often, newcomers end up in
crowded shantytowns, favelas, and slums, because cities concentrate
the very poverty from which they offer an escape. But that won't
slow the influx as long as hope wears Nikes and is steeped in fumes.

Oz-like cities shimmer as beacons of prosperity, with enhanced
education, better medicine, more jobs for women, and wide streaks
of upward mobility. Even environmentally, cities can eclipse sparsely

settled country life. When roads, power lines, and sewers lie closer together, they require fewer resources. Apartments are insulated by the civil geometry of the buildings, making them easier to heat, cool, and light. Crowded neighbors can share public transportation, and most destinations tend to be close, within walking or biking distance; people rarely need cars. As a result, city-dwellers actually create a much smaller carbon footprint than rural-dwellers do. Cities like New York boast the lowest amount of energy use per household and per person, and so, paradoxically, although the city as a whole uses more energy, each person uses less. It seems counterintuitive, but city life can be a more eco-friendly way for humans to live. Cities in developing countries also use less energy—but that's because the number of poor tends to be higher there and they consume less, including less food and fresh water.

Still, despite clustering services and leaving smaller individual footprints, the record number of people fleeing the countryside for city life is worrisome to climatologists, because cities are environmental game-changers. Big cities are hotspots, on average ten degrees warmer than their surroundings, and they emit most of the planet's pollution, as cars prowl their streets and food caravans travel long distances to stock their groceries. On some summer days, the air hangs thickly visible, like the combined exhalations of millions of souls. Steam rising from vents underground makes you wonder if there isn't one giant sweat gland lodged beneath the city.

One of the paradoxes of our age is that we're urban primates who are still adapted to the wilderness, which we long for and need, at the same time that we're destroying, building over, and farming all that's wild. Since the crowd-rush to these asphalt jungles is accelerating, we need ingenious ways of harmonizing city life with human and planetary well-being. Our challenge will be finding a way to have both, while also preserving the planet.

Some of the best ideas I've encountered do just that, transforming our cities from grimy energy guzzlers into dynamic ecosystems.

City parks are essential, but in addition, picture shade-loving wildflowers in gated alleyways, fresh vegetables growing on roofs and piers, lushly planted walls, vertical farms in skyscrapers, rooftop beehives brewing honey, and nature trails threading through rusty old infrastructure. Greening a city with vegetation is a proven way to cool it down, filter the air, suck out carbon dioxide, ladle in more oxygen, and offer pockets of calm amid the bustle and din.

Hoping to achieve that intermingling, a new branch of environmentalism known as "Reconciliation Ecology" has emerged, which strives to preserve biodiversity on our doorstep in cities and other human-dominated habitats. The term "reconciliation ecology" was coined by Michael Rosenzweig in his book *Win-Win Ecology*, and it has a lovely ring to it. It suggests fence-mending and coexisting in harmony, not a wallop of blame. It's based on figures showing that we haven't enough unsettled land left on Earth to protect all of life's biodiversity, but we can make room for plenty more in our cities and yards.

Along country roads near my house, where cornfields and houses predominate, you'll see nest boxes for bluebirds, provided by thoughtful bird-lovers because natural tree cavities have grown scarce. When replacing wooden fence posts with steel ones led to the rapid disappearance of shrikes (medium-sized birds with hooked beaks like birds of prey), locals restored the wooden fence posts (on which shrikes like to perch), and the shrikes returned. These may be small acts of reconciliation, but if you create enough of them it can change the big picture. And not just in the countryside. Some of the most improbable-sounding efforts are blurring the line between civilized and wild. "Wastewater Treatment Plant" may not sound like a natural or particularly scenic destination. But some symbiotically minded towns have been designing a new breed of wildlife preserve, one that gives recycling a lively twist. Instead of dumping treated water, they return it to nature as the essence of an ecosystem that offers food and habitat to animals. As the water is further purified by vegetation, migrating and native birds find a home, entangled communities of plants and insects take up residence, and a hodgepodge of wild animals bustle in.

City-dwellers then needn't travel far for an interlude to refresh their habit-dulled senses. Strolling, gawking, sitting, camera-clicking, humans become one more changing feature in the perpetual tableau, another flock of familiar creatures to whom the scores of nesting birds pay little mind.

A favorite such preserve of mine is the Wakodahatchee Wetlands in suburban Delray Beach, Florida. On a boardwalk raised ten feet above any hazard, one can watch an alligator gliding among the bulrushes, fish defending their mud nests from marauding turtles, dabbling ducks and teals, wading birds stalking their prey. The shallow water and raised trail make many dramas visible from above, including whiskered otters catching whiskered sail-finned catfish. Pig frogs grunt like their namesakes. If you're lucky, you might see a patch of water fizzing like frying diamonds in the sun—where a male gator is bellowing in a bass too low for human ears. Or you might spot a giant prehistoric apparition standing like a sentinel in the water, as an endangered wood stork displays its distinctive gnarled-wood bald head and long curved beak.

The wooden walkway loops for nearly a mile through fifty acres of swamps, marshes, ponds, reeds, and bogs. Wherever water and land meet, life seems to thrive. Flapping around an archipelago of brushy and treed islands, ibis run a regular feeding patrol to nests full of squawking chicks.

Despite the usual urban hubbub, 140 species of birds broadcast on every channel at Wakodahatchee, from a pitying of collared doves to a pandemonium of monk parakeets. Red-nosed moorhens create a steady background score of trumpeting, clucking, and loud monkey-cackling. Courting roseate spoonbills play the castanets of their bills. Red-winged blackbirds spout the only buzzwords.

Though surrounded by restaurants, offices, condos, malls, and highways, Wakodahatchee's wetlands attract a bounty of life, including wild plants one rarely sees in cities. What at first seems a flush of algae, or a pointillist canvas of sunstruck water, is brilliant chartreuse duckweed. This simple aquatic plant floats everywhere on the slower-moving waters of our planet, offering food to birds, shade to frogs and fish, and a warm blanket to alligators and small fry. One day it may also provide a cheap source of high protein for humans (it's already eaten as a vegetable in some parts of Asia) or a cheap producer of biofuel that will power cars while filtering carbon dioxide from the air.

There's no stigma attached to reconciliation projects being lucrative. Israel's Red Sea Star Restaurant, for example, 230 feet off the shore of Eilat, is a combination bistro and observatory, seating people in its colorful, marine-inspired dining room sixteen feet down from the surface on the sandy sea floor. Plexiglas windows offer diners, sitting on squid-shaped chairs under dimmed, anemone-shaped lights, a view of a wealth of sea creatures in the coral gardens by day or night. Equally curious fish also get to ogle the diners. It happens to be an architectural showpiece, but it's also an ecological triumph that has restored a coral reef that was lost through human pollution and overuse. Architects began by choosing a barren stretch of sea floor, laying down an iron meshwork, and transplanting coral colonies onto the trellis, where they cling like slow-motion trapeze artists and continue to attract marine life.

In an acrylic tube submerged in the Indian Ocean, at Hilton Hotels' Ithaa Restaurant, in the Maldives, diners are also surrounded by fish and coral as they eat. Although the Maldives, a nation of islands only five feet above sea level, emits but a tiny fraction of the world's pollution, its president, Mohamed Nasheed, has set the most ambitious climate goals of any country on Earth, promising to go carbon neutral within ten years, while building sustainable hotels and restaurants and even a floating golf course. "Our oil-fired power stations will be replaced with solar, wind, and biomass plants," Nasheed explains. "Our waste will be turned into clean electricity through pyrolysis technology, and a new generation of boats will slash marine transport pollution. By 2020, the use of fossil fuels will be virtually eliminated in the Maldivian archipelago." Greening the economy is good for the Maldives, which has begun attracting a flock of eco-tourists and investors, and it's also a model for changes radical enough to help fix the climate.

In some cities, coexisting with nature means salvaging rusty old infrastructure, reclaiming abandoned blocks and trashyards, and forging junked metal into inviting habitats for plants, animals, and humans. In every U.S. state, and many other countries from Iceland to Estonia, Australia, and Peru, out-of-work railroad tracks have morphed into peaceful "rail trails" ideal for biking, hiking, and cross-country skiing. Most often they slip through towns or skirt farmlands, drawing both humans and wildlife to leafy byways. I've biked or cross-country skied on some beauties in Ohio, California, Arizona, and New York. Memorably, biking on a rail trail outside Gambier, Ohio, at dawn, I was chased by a flock of farm geese. I knew their charge was merely bravado, so I pedaled slowly and let them nip at my pant legs, which seemed to give them a sense of territorial satisfaction, and they soon returned to the barnyard. I enjoyed their brief honking companionship, and learned something about geese I didn't know: what clamorous watchdogs they make.

A different stripe of oasis growing in popularity is the High Line on Manhattan's West Side, a surprising sprawl of undulating benches, nests, perches, and lookouts, giving New York City yet another bridge—this one between the urban and the rural. An old elevated freight spur, little more than a rusty eyesore on the Hudson, it's metamorphosed into a tapestry of self-seeding wildflowers and domestic blooms. It isn't the first raised park (there's the Promenade Plantée in Paris, and remember the Hanging Gardens of Babylon?), but the High Line is the loveliest city rail trail I know.

Picturesque, with many scenic views, it's also richly detailed and alive, allowing you to feel elevated in spirit, floating in a garden in space where butterflies, birds, humans, and other organisms mingle. In a practical sense, it's a lofty shortcut, a sky alley that avoids all the intersections. A million people have already strolled its land-scaped corridors, and it's inspired other cities hoping for similar sky parks. Chicago, Mexico City, Rotterdam, Santiago, and Jerusalem are among those following suit with their derelict trestles, each an urban renewal project featuring regional plants and its own special character or sense of humor. In Wuppertal, Germany, the rails-to-trails corridor includes a brightly colored LEGO-style bridge. Like the wastewater wetlands, such projects are widening our notion of recycling and yielding an urban lifestyle that's interwoven with nature.

As one salve in our medicine cabinet of good ideas, these vest-pocket urban parks and wildlife corridors have deep roots around the world, from nest boxes for storks in Romania, Switzerland, Poland, Germany, Spain, and other havens along their well-flapped migration routes to species-rich Central Park in the heart of New York City, London's eight city parks (in several of which deer roam), the temperate rainforest of Vancouver's Stanley Park, Moscow's Losiny Ostrov ("Moose Island") National Park, Sanjay Gandhi National Park in Mumbai, sod roofs and greenways from Germany to the Faroe Islands, St. Luke's Hospital rooftop garden in Tokyo, mandatory rooftop gardens in Copenhagen (and proposed in Toronto). A surprising jumble of native species thrives in and around heavily planted Rio de Janeiro, Istanbul, Cape Town, Stockholm, and Chicago, which have become biodiversity hotspots. Then there's Singapore's big, blooming, downtown Gardens by the Bay, enriching city life with more than 240,000 rare plants, flowers, and trees in domes that rise sixteen stories over the city. Including a cloud forest and aerial walkways, the gardens collect rainwater, generate solar electricity, and bathe the air. Opening on June 29, 2012, they drew 70,000 nature-hungry visitors during the first two days.

Although these new city oases won't work for all species, or for all communities, the trend for rewilding our cities is growing. It's positive, it enlightens, it's widespread, and it helps. We need to retrofit and reimagine cities as planet-friendly citadels. They're our hives and reefs. Sea mussels aren't the only animals living in individual shells that are glued together.

A GREEN MAN
IN A GREEN SHADE

As a child, Patrick Blanc loved going to the doctor's office. A six-foot-long aquarium in the waiting room tickled his eyes with colorful tropical fish and plush green plants that swayed in the current and beckoned like hands. To an urban boy growing up in the Parisian suburbs, it offered a glimpse of paradise. When he put his ear to the small box attached to the aquarium, he heard the bubbling of water through tubes and the filter's hum. The hydraulics and engineering intrigued him as much as the fish, and before long he designed his own small aquarium at home. For a spell, he also adopted coral-beaked waxbills, setting them free on Thursday and Sunday mornings to fly around the apartment.

As he entered adolescence, his nomadic curiosity drifted from aquariums and birds to aquatic plants, and then, at fifteen, he leapt above the waves to the world's moist shaded zones, the mysterious understory of tropical forests. A college trip to the rainforests of Thailand and Malaysia brought the revelation that "plants could sprout at any height, not merely from the ground."

Now a bounty of planted walls refreshing cities around the world owe their design or inspiration to Blanc's eco-pageants, and lure birds, butterflies, and hummingbirds while they mutate with the seasons.

One of Blanc's personal favorites, a green city icon, is the magnificent Quai Branly Museum in Paris, which opened in 2006 and was greeted by many as a botanical epiphany. Multitextured meadows climb the thirteen-thousand-square-foot facade of the building, more than half of which is alive. The rest is windows, creating a giant plaid of thick-leafed, mossy, breathing wall, touchably soft, rich with scent, atwitch with birds.

Cloaking the facade in great variety to reflect the cultural diversity of the world's artists, Blanc chose a pastiche of species from temperate zones in North America, Europe, South America, Asia, and Africa. He would have included Oceania, but tropical plants can't survive the Paris winters, and the facade—which is part botanical tapestry, part hidden lagoon, and entirely soil-free—is intended to endure for many years, filling the senses of Parisians, and building a 40-foot-tall, 650-foot-wide ecosystem amid the hard premises of city life, while also helping to purify the air and eliminate carbon dioxide. In warm weather, flowers bloom, butterflies nectar, and birds perch and nest in the dense thickets. One half-expects to find miniature deer browsing on its mossy hillocks. The museum director plans on adding frogs and tree lizards. Because our horizontal indoor life can flatten the mind, some of the administrative offices also have smaller vertical gardens, which are visible through the windows, blurring the line between outside and inside even more.

How can a towering garden with a northern exposure survive the icy winds sweeping across the Seine? That's where Blanc's education and research as a botanist come in. The living wall is hardy because he's chosen hundreds of understory species that he's discovered can nonetheless tolerate slathers of direct light and wind.

"When I think of *Heuchera*," he says, referring to a family of plants that includes coral bells and alumroot, which produce small delicate flowers, and whose leaves are like hands with the fingers extended,

"I always think of their leaves emerging intact from the melting snow in April, along the steep slopes in the shade of giant sequoias in California."

Blanc works with a palette of deep rich greens in dozens of subtle shades and intensities, from asparagus and fern green to forest or praying mantis green, and textures that run the gamut from matte to hairy, spongy to sheen. All vary with time of day, age, season, clouds eclipsing the sun, fog rolling off the river, rush-hour traffic, aberrations of twilight. Seen through our rods and cones, the colors remix and evolve perpetually as they would if we encountered them in a forest. He prefers leaves to flowers, doesn't care for trailing vines, and is sensitive to the architecture of leaves. The thousands of individual plants he quilts together grow leaves that are bristled, pointed, star-shaped, notched, oval, sickle-shaped, circular, teardrop, blunt, heart-shaped, arrow-headed, and more. Some climb while others descend, some mound or bloom daintily, others sprout or cantilever. Knowing the habit of each, he draws a multicelled planting map that looks like swirling fingerprints or a paint-by-number guide, with each segment a plant species referred to by Latin name.

"They begin like paintings," he explains. "Then they develop texture and depth."

As a science-based art form, it's a fusion inspired by many muses. The plants are drawn on flat paper, so each design does indeed begin like a painting. Then the artwork morphs into a sensuous sculpture of touchable, biological, prunable shapes and colors. Leaves, flowers, stems dance in the air, a slow-motion ballet. He may try to choreograph them to some degree, but the ensemble will succumb to wild swings of improvisation, depending on the weather. As frogs, birds, and insects take up residence, they'll add a croaking-chirping-buzzing chorale, with some notes foreseeable and the rest jazz variations.

Although the plants naturally curve, clear lines and clean edges give the finished work a tone of sensuous elegance, not disarray. It's lofty and complex, not cluttered. The plants aren't exactly wild, but they can flourish in unique ways. In that sense, it's more like chaos aligned—a deliberate, contained, carefully measured, masterfully executed free-for-all.

Practical botany, hydraulics, physics, and materials science are essential scaffolding for the artistry. Thousands of individual plants are inserted by hand into pockets on a flat felt sheet that's rigidly framed and will be watered and fed, rain-style, by intermittent showers from a hidden pipe running across the top. Despite the lack of soil, the plants quickly flourish, covering the felt and pipe. The overall effect is a gulp of wild nature that hits you in the solar plexus. This is a garden you stand up and greet at eye level, as you would a person. It invites you to touch and smell it. Look up, and it looms four stories above you like an expansive forest understory, not a fairy-tale giant. Close in, it creates its own weather bubble and is quite shady and moist if you stand beside it and tilt up your chin at the dizzying vegetation. Balancing on a very thin wire between tame and free-willed, it seems both intimate and indomitable.

VERTICAL GARDENS, LIVING roofs, and urban farms are going mainstream everywhere. A few of my favorites: Mexico City's towering arches carpeted in fifty thousand plants astride car-clogged avenues; the blooming brocade of native plants adorning the inner walls of the Dolce Vita shopping center in Lisbon; the glassed-in courtyard of Milan's Café Trussardi, where a canopy of frizzy greens and purples floats above diners and cocktail-sipping flaneurs, trailing vines and flowers like a hint of heaven; the golden wheatfield atop the Canadian War Museum in Ottawa; nine sinuous houses buried under earth and grass in Dietikon, Switzerland; the Grange, atop two buildings in the Brooklyn Navy Yard, where you're surrounded by organic vegetables and views of the Hudson. The roof of the Chicago Botanic Gardens' Rice Plant Conservation Science Center doubles as a garden visited by millions of people and a botanical laboratory, and the roof of Chicago's City Hall also serves as a study site, with the usual black tar on one side and a wildflower garden on the other. (On summer days, the ambient air above the planted side measures as much as 78°F cooler than the air over the old blacktop.) And living rooftops are becoming hot property. One U.S. company has already sold 1.2 million square feet of sprouting, blooming, bird-, bee-, and butterfly-enticing roofs, mainly to private residences.

Other living roofs and vertical garden companies have sprung up and begun greening all sorts of buildings, from hospitals, homes, and police stations to banks and offices. Some use hydroponic methods; others plant in turf on the roof à la the European tradition. A former Renault factory in Boulogne, France, has been reborn as a school with an undulating green-roof that reduces heating and cooling costs. The design firm Green over Grey, based in Vancouver, British Columbia, has created some spectacular living walls for sites in Canada, including the international building at Edmonton Airport, where arriving visitors inhale a shot of oxygen and plant-scrubbed fresh air from a gigantic living wall whose design swirls were inspired by high-altitude cloud formations. "Jungle Waterfall," their dramatic, multistory cascade in an office building in Vancouver, includes tropical trees, and a maintenance crew occasionally has to harvest the pineapples lest they fall and hit passersby.

Planted walls and roofs and sustainably designed buildings, along with wildlife corridors, city parks, solar and wind power, and trees aglow with bioluminescent foliage (to replace streetlamps), are but a few of the initiatives gaining popularity in the U.K., Germany, Taiwan, the United States, and many other places worldwide. Roofs planted with sedums and succulents blossom, changing color with the seasons, while being low-maintenance, reflecting heat, and providing a habitat for birds. The goal is homes and public spaces that are living organisms that will scrub the air of pollutants, increase oxygen, reduce noise, save energy, refresh the spirit, and sink our roots deeper into the natural world.

———

IN CONTRAST TO Blanc's elegantly formal walls for the Quai Branly and a similar project at the Athenaeum Hotel in London, his own house on the outskirts of Paris is a throbbing green Mardi Gras of microhabitats, with knolls of lance-shaped leaves, jutting rocks, prongs of tiny flowers, thick heart-shaped leaves, arrow-leafed philo-dendrons with roots adrift in a flowing stream. You don't so much enter the abode he shares with his longtime partner, the actor Pascal Henri (a.k.a. "Pascal of Bollywood"), as join a green rhapsody, or possibly a green bedlam, and become part of the cascading carnival of fronds, mounds of moss, umbrellaing ferns, twig elbows, search-ing roots, and leafy limbs probing at everything, including you. The superabundance of leaves caresses you lightly with barely discern-ible veined fingers as you pass. Beaded curtains serve as doors, and free-flying coral-beaked waxbills wing from room to room. Bounc-ing frogs and slithering lizards roam the house at ease, eyes rotating, tongues occasionally unfurling like party favors. At a Magritte-like Surrealist window, a bushy shrub on the inside echoes its twin on the outside. Your eye shimmies. Inside, outside—who can say where they begin or end? Glass is only liquid sand, after all, and only ever in motion, hourglasslike, pouring so slowly that our eyes read it as solid.

Blanc's study is a green thought in a green shade where he's walk-ing on water. Literally. The floor is a sheet of plate glass atop a 20' x 23' aquarium, home to a lush expanse of vegetation and over a thou-sand tropical fish. Holding 5,283 gallons of water, it's loaded with plants whose long white roots ripple like medusas, as they naturally purify the water and also provide grottos for the fish.

The large wall behind his glasswork table is a weave of plants with plush textures, mossy tussocks, cascading fronds, and a kaleidoscope of greens. At the base of the living walls, a narrow stream flows, pro-viding nourishment for roots and refreshment and nesting sites for the birds. He dips a hand in as a bird glides overhead to perch in the rhododendrons. Here and there, algae, moss, and liverworts have sprung up on their own. In a large bookcase, nearly all the books have green jackets. Only little brown bats and bombardier beetles are missing.

"I take my shower outside every day, even if it's snowing," he confesses. "I refuse to heed the limits between inside and outside imposed on a human lifestyle that migrated from tropical origins to colder, even glacial climates. To heighten the absurdity, life in tropical cities requires air-conditioning to cool the indoor atmosphere. Wherever one goes in the world, regardless of the season, it is necessary to either heat or cool dwellings." This merits a hand lifted to the absurdity. "We need buildings with a better thermal balance."

Despite his serious purpose, a spell of playfulness pervades his preferred habitat, as it does his person. His shirts all seem to be patterned in leaf designs; he wears green shoes, has a two-inch-long thumbnail painted forest green, and wears a streak of bright green in his hair. For a moment I think he might have a single leaf of *Iris japonica* growing from his skull—his green forelock is shaped like one of its long, tapering leaves. It is his signature plant: one often sees *Iris japonica* dangling down forest edges in the wild. Blanc uses it in most of his installations as an echo of gently cascading water.

"We live in an era where human activity is overwhelming," he continues. His chilled white wine, Vogue menthol cigarette, computer, and electric lights make it clear that he does appreciate cosmopolitan life. In fact, he's spent all of his life living in cities—while making forays to some of the wildest places on Earth.

"I think we can reconcile nature and man to a much greater degree."

He's not alone in that conviction. A good start may be rethinking our houses, because at the racing heart of every city is still the ancient, unalienable idea of home.

HOUSE PLANTS? HOW PASSÉ

Home, for the Inuit, had an elemental simplicity. They used bone knives to carve bricks from quarries of hardened snow. A short, low tunnel led to the front door, trapping heat in and fierce cold and critters out. Mortar wasn't needed, because the snow bricks were shaved to fit, and at night the dome ossified into a glistening ice fort, with the human warmth inside melting the ice just enough to seal the seams. The idea behind such homes was refuge from elements and predators, based on a watchful understanding of both. The igloo was really an extension of the self—shoulder blades of snow and backbone of ice, beneath which a family slept, swathed in thick animal fur, beside one or two small blubber lamps. All the building materials lay at hand, perpetually recycled, costing nothing but effort.

Picture most of our houses and apartment buildings today—full of sharp angles, lit by bulbs and colors one doesn't find in nature, built from plywood, linoleum, iron, cement, and glass. Despite their style, efficiency, and maybe good location, they don't always offer us a sense of sanctuary, rest, or well-being. And they're not particularly healthy. A U.S. Environmental Protection Agency study found levels of twelve volatile compounds two to five times higher indoors—no matter if the home was rural or urban—due to the products we use and poor ventilation. Because we can't escape our ancient hunger to live close to nature, we instinctively encircle the house with lawns and gardens, install picture windows, adopt pets and Boston ferns, and scent everything that touches our lives.

No wonder there's an impassioned push worldwide to build green homes with verdant walls and roofs, inspired by Patrick Blanc, equally green workplaces that breathe and clean themselves like street cats, and well-tilled farms on rooftops and in ziggurats. It doesn't make sense to shut out nature in the old way. Our fundamental archetype of a foursquare, armorlike building perched on a scrap of earth is evolving from a static and ultimately disposable dwelling into one that, like a tree, mingles holistically with the world around it, not just absorbing a staggering amount of nutrients but producing even more than it consumes.

An alternative is the culture of sustainability and "cradle to cradle" design redefining the world of goods and architecture and city planning. According to the principle of "cradle to cradle" (a term coined by the Swiss architect Walter R. Stahel in the 1970s), everything we make—apartment buildings, bridges, toys, clothes—should be designed with reclamation and rebirth in mind. Instead of tossing the outmoded ephemera of civilization onto rubbish heaps, and then extracting and grinding down more resources to replace them, why not fabricate objects that will naturally biodegrade or can be recycled by industry as "technical nutrients"? Durables such as televisions, cars, computers, refrigerators, heaters, and carpets could be leased and traded in when worn out or untrendy, allowing manufacturers to recycle them and harvest the raw materials.

In 1999 the architect William McDonough accepted the challenge of redesigning Ford Motor Company's eighty-five-year-old River Rouge factory, a project that required redesigning the ten-acre roof of its 1.1-million-square-foot truck assembly plant. He began by endowing the roof with its own weather system—acres of sedum, a low-growing succulent that blooms dusty-pink or linen-white in the fall and the rest of the year displays large rain-swollen leaves. Then he knitted the factory and plants into the landscape with "a system of wet meadow gardens, porous paving, hedgerows and bio-swales that attenuates, cleanses, and conveys storm water across the site."

Inspired by such models, and hoping to rank high on the prestigious LEED (Leadership in Energy and Environmental Design) rating system, architects are vying to create equally well-behaved buildings that "are environmentally responsible, profitable and healthy places to live and work." They're striving for regenerative buildings that purify their wastewater, create more energy than they use, and compost and recycle to such an extent that industry blends seamlessly with nature. "In essence," Andres Edwards writes in *The Sustainability Revolution*, "a world of abundance, rather than limits, pollution, and waste." This revolution stems from an ethos that's reverberating around the world in developed and developing countries alike; as Edwards reminds us, "Brazil, Canada, China, Guatemala, India, Italy, Japan, Mexico, and Netherlands Antilles have LEED-registered projects, demonstrating that the standard can adapt to different cultures and bioregions."

We aren't adhering any longer to the myth that food must be grown far away and transported on trucks. We can easily envisage restaurant rooftop farms, urban beekeeping and midtown chicken coops. We can't grow everything around the corner—not grains, or soy, or corn, to be sure—but we can grow most of our vegetables and fruits. Local farms feed the food chain, save fuel, and guarantee a fresher and more nutritious diet. And they're cropping up on every continent, including the last place one might guess.

In Antarctica, where the average coastal temperature is -70°F with inland dips to -180°F, the American research base, McMurdo Station, is a town of naked machines and heavily insulated people. There darkness saturates winter, inking out the sky for six months, during which occasional green auroras shoot up like magnetic demons' tails, and indoors the auroras are falling white showers of man-made fluorescence. There are two primary smells (sweat and diesel fuel) and two primary colors (black and white). Fresh produce arrives by air from Los Angeles and costs $80,000 to $100,000 per week in the summer. During the winter, deliveries may be months apart.

"Clearly, this is no banana belt," says Robert Taylor, the good-humored technician who, along with many volunteers, has overseen McMurdo's 649-square-foot greenhouse. "And there is no history of oxen tied to wooden plows turning over rich black soil. Actually, on this side of the continent, there is no soil at all, only weathered volcanic rock and, of course, ice. There is nothing in the way of organic matter to speak of, and no recognizable terrestrial plants. And yet, life blooms . . . under thousands of watts of artificial light."

It's hardly roomy, especially compared to gardens in his hometown of Missoula, Montana. But by using hydroponic techniques he's been able to harvest about 3,600 pounds of spinach, Swiss chard, cucumbers, herbs, tomatoes, peppers, and other vegetables each year—pure manna to the green-starved residents.

"Not enough to register on the world's export market, but nothing to sneer at if you are one of the approximately 200 people who choose to winter here," Taylor says via e-mail.

"Lettuces grow like champs," he notes. "There are nearly 900 lettuce heads growing at any time on tiered growing systems. Likewise, basil and parsley are herbs that need very little in the way of input." That's just as well, because he has to pollinate them all by hand, since insects, the natural pollinators, are forbidden, lest they devastate the small greenhouse Eden.

"It's strange that a horticulturist would come all the way to Antarctica to grow vegetables, but as far as challenges and thrills, what better place to confront the beauty of plants than in an environment so devoid of them? . . . Each tomato, each cucumber becomes a jewel, precious."

As if it were a laid-back bar in Key West, two hammocks and a cozy old armchair float in a humid corner, "for those who wish to commune with arugula." Many do. At McMurdo, not only vegetation but humidity, scent, and natural colors are rare. On the other hand, howling isolation and intense relationships are the norm. Many people thrive on the parabolic sunlight and unusually intimate community. But those beset by "polar T3," overwintering syndrome, can find their thyroid levels askew and metabolism rocky, with sleeplessness, irritation, and depression constant bedfellows. In an all-white kingdom of ice and snow, where the only low-hanging fruit are the stars, one's sanity can tremble on a stem slender as a marigold's.

Fortunately, purple-and-yellow pansies and orange marigolds (both edible) grow in the greenhouse, where the rainbow stalks of Swiss chard create a small psychedelic forest, and scarlet cherry tomatoes dangle from string supports like floppy marionettes. Cilantro, basil, chives, rosemary, and thyme scent the air. The sensory repast as well as the food nourishes greenhouse visitors, and the plants lap up the CO_2 exhaled by the humans. Unlike typical greenhouses in winter, this one has no sunlight streaming through cathedral-like walls of glass. McMurdo's urban farm at the bottom of the world is completely sealed and insulated, and, in a stark village where windows are precious, I'm told it also offers a leafy idyll for a dinner date. Even in this extreme outpost of a city, the benefits of greening ease the way.

There will soon come a time when farming needn't have a country flavor, and referring to "the north forty" means crops forty floors up. PlantLab in the Netherlands grows forty different crops indoors, using hydroponics and high-tech sensors, without pesticides, and even without windows. Plants don't need the whole spectrum of light; instead each crop is raised with the precise amount of blue or red light it craves. As the water evaporates, it's recycled, so only a pittance extra is needed. In these specially controlled environments, the crop yield is three times higher than outdoors, and the process would do equally well in the Sahara or Siberia once LED lights become a bit cheaper.

All of our buildings need to earn their keep. We're probably in the last era of deadbeat buildings. In the United States alone, buildings use 40 percent of the country's raw materials, burn 65 percent of the total electricity, and drain 12 percent of drinkable water, while piling up 136 million tons a year of demolition and construction wastes.

SUPPOSE THE GOAL is buildings that are inherently living organisms. Just how alive could a home or office become? In addition to living walls and planted roofs, its skin could mimic plant metabolism and animal musculature. "Biomimicry" is an old idea but a dynamic and lucrative new direction in architecture and engineering that mines the genius of nature to find sustainable solutions to knotty human problems.

Picture: Houses painted with lotusin, a self-cleaning paint inspired by the water-shedding veneer of leaves. Products colored without pigments, echoing the way light dances across peacock and blue-jay feathers. New lenses and fiber optics that mimic the almost distortion-free lenses coating the body of a brittle sea star, or the flexible optics atop a sea sponge's tentacles. Electronic devices inspired by mussel tissue, which automatically dissolve when you discard them. A building whose outer skin resembles the porelike stomata of leaves and provides all the energy it needs. Ships' hulls engineered like whale skin to glide through the water while burning less fuel, and airplane wings that save fuel by mimicking ripple-edged whale fins.

The result is organic, self-assembling, nonpolluting solutions that nature has already mastered and we can copy. This frame of mind requires a major flip in our way of thinking and our sense of how we exist in nature. For the longest time, "heat, beat, and treat" was the industrial motto. We've built cities and fueled empires by raiding the Earth's resources, chopping them up, heating them, breaking them down with toxic chemicals, and fastening them together. Biomimicry asks: "Okay, that's how humans make things—and it doesn't work. How does *life* make things?"

"Organisms have figured out ways to do the miraculous things they do," the biomimicry pioneer Janine Benyus says, "without jeopardizing the future of their resources and offspring."

Inspired by Benyus and others, cities are blooming with architecture that functions like (and sometimes resembles) growing organisms. Imagine transparent skyscrapers that save energy as their facades expand and contract like an elaborate array of muscles. In a working prototype designed by the New York firm Decker Yeadon, swirling silver ribbons in the glass facade are really a three-layered muscle: a rubbery polymer sheathing a flexible polymer core, with a silver coating that skittles an electrical charge across the surface. If it's too cold, the ribbon-muscles "fire" and contract to slender squiggles, exposing lots of window to the sun. In hot weather, the ribbons expand like a patchwork of shot silk to create a flat parasol of shade. Many small segments self-regulate in this way, fiddling with their own thermostat to stay in homeostasis. Much as we do. Too warm? Shed the sweater and move out of the direct sun. As a design, muscular walls are more flexible and stronger than solar panels.

Or picture the high-rise office and shopping complex Eastgate Centre in Harare, Zimbabwe, which was inspired by a throng of gigantic termite towers. Topped by turrets and pinnacles, a city of vaulted termite mounds can rise thirty feet from the parched earth like otherworldly castles, while millions of laborers and soldiers toil inside, presided over by king and queen, with offspring raised communally. It's an agricultural society, whose favorite crop is a fungus that will only grow at 87°F. Yet outside the thick mud walls, temperatures can plunge to near freezing at night and soar to broiling by day. Winds snort and snap one moment and tap like weary ghosts the next. Our windmills and wind turbines require a steady flow of wind, and they're stymied by turbulence. Termite engineers harness chaotic winds far more skillfully by using their mounds as inside-out lungs.

Deep inside each mountainous city, termites capture and trim even the most sputtering, heat-whipped, muddled winds to the precise vibration needed to ventilate the colony and keep their crops flourishing. As they open and close a filigree of low doors, the mound inhales a rush of air into a maze of chambers and passageways, and shoots it up to the buttresses and tiptop chimneys. They keep tweaking their design by opening and closing doors, digging new doors, sealing old ones, adding wet mud in spots for quicker cooling. Each termite is like one neuron in a collective brain. It doesn't need to be smart, and none can see the whole picture, but together they create coherent action and a kind of intelligence. Constant gardeners, they fine-tune the breeze and provide a steady temperature, which keeps the tiny, blind population cozy.

We have this in common: they are great openers of doors, as we are, though their doors are physical and many of ours are symbolic. Some of their blade-shaped "compass" mounds are oriented toward the sun at a time-tested angle to avoid the roasting noonday sun, yet usher in evening's faint rays. My house was designed on the same principle by its first owner—vaulted south-facing windows in the living room let in more winter sun and provide summer shade.

Singing can draw oxygen through even injured lungs, and the mud colony fills with a vibrato, reassuring its lodgers that all is well. On some level unknown to us, an out-of-tune mound must not sound right to the termites, who are driven to build the most exquisite lungs possible because their lives depend on it. Does this mean they have an aesthetics that's shared by the whole colony? Who's to say. It could be that, to them, out-of-tune air stings like a million arrows.

Inspired by mud casts of the mound's baroque nooks and crannies, the African architect Mick Pearce designed the Eastgate Centre. Fans on the first floor spirit air through ducts into the central spine of the building, and stale air seeps through exhaust vents on each floor, exiting at last from high chimneys. A river of fresh air automatically replaces it. Using only 10 percent as much energy as nearby buildings, the eco-friendly Centre has saved the owners $3.5 million in climate control, which they've passed on to their tenants by lowering the rent. Ten years later, Pearce built the even more efficient ten-story Council House 2 building in Melbourne, Australia. This time, recycled wooden shutters, covering one whole side of the building, open like petals at night to expel the warm air from offices and shops. This works well in Africa, but in colder climes excess warmth can't be wasted. And so, in some countries, furnaces now have two legs.

OPPORTUNITY WARMS

How intimate, how romantic, how sustainable of the French. As I waited with a throng of Parisians in Paris's Rambuteau subway station on a blustery November day, my frozen toes finally began to thaw. Alone we may have shivered, but together we brewed so much body heat that people began unbuttoning their dark coats. We might have been emperor penguins crowding for warmth in Antarctica's icy torment of winds.

Idly mingling, a human body radiates about 100 watts of excess heat, which can add up fast in confined spaces. Rushing commuters contribute even more, and heat also looms from the friction of trains on the tracks and seeps from the deep maze of tunnels, raising the platform temperature to around 70°F, almost a geothermal spa. As new people clambered on and off trains, and trickled up and down the staircases to Rue Beaubourg, their haste kept the communal den toasty.

Geothermal warmth may abound in volcanic Iceland, but it's not easy to come by in downtown Paris. So why waste it? Instead mine people as a renewable green energy source. Tap even a fraction of the population, say, the heat cloud of subway commuters, and it's a deep pocketful of free energy. In that spirit, savvy architects from Paris Habitat decided to borrow the surplus energy from all the hurrying bodies in the metro station and convert it into radiant underfloor heating for apartments in a nearby social housing project, which happens to share an unused stairwell with the station. Otherwise the heat borne through countless rushed breakfasts of croissant and café au lait, mind-theaters, idle reveries, and flights of boredom would be lost by the end of the morning rush hour. Opportunity warms.

Appealing as the design may be, it isn't feasible throughout Paris without a pricey retrofit of buildings and metro stops. But it's proving successful elsewhere. In America, there's Minnesota's prairielike monument to capitalism, the four-million-square-foot Mall of America. Even on subzero winter days the indoor temperature skirts 70°F from combined body heat, light fixtures, and sunlight cascading in through 1.2 miles of skylights. That's just as well, since people can get married in the mall's Chapel of Love on the third floor, next to Bloomingdale's, and taffeta and chiffon aren't the best insulators.

Or consider Scandinavia's busiest travel hub, Stockholm's Central Station, during morning rush hour on a blustery day in January. Outside, it's -7°F, the streets are icy as a toboggan run, cold squirrels around your face, the air feels scratchy, and even in wool mittens your hands are tusks of ice. But indoors is another country, a temperate one filled with a living mass of humans heading in all directions. Pocketing the windfall, engineers are harnessing the body heat issuing from 250,000 railway travelers to help warm the thirteen-story Kungsbrohuset office building about a hundred yards away. Under the voluminous roof of the station, travelers donate their 100 watts of surplus natural heat, while visitors bustle around the dozens of shops, buying meals, drinks, books, flowers, cosmetics, and such, bestowing even more energy.

You can almost feel a gentle tugging at your skin. There's a warm draft.

"Why shouldn't we use it?" asks Klas Johansson, who works for

Jernhusen, the state-owned developer of the project. "If we don't use it it's just going to be ventilated away."

Citizen lamplighters, citizen furnace-stokers—antique corps of volunteers fill my imagination. All cheap and renewable. This ultra-green design works dramatically well in Sweden, a land of soaring fuel costs, ecologically minded citizens, and a legendary arctic winter becalmed by few hours of daylight and a horizon-hugging sun. Night blankets the city in stellar darkness by midafternoon. Offset as the night may be with cozy lantern-lit streets, candle-lit windows, and the luminous green ribbons and dancing halos of the northern lights, when cold clambers up the bones, more heat is the sole comfort. But fuel can take many forms, from fossil to solar energy, oil and gas to the residue left in paper mills, or Central Station's . . . well, what shall we call it? As a technology it needs a catchy name, something sociable. Maybe "Auraglow," "Beradiant," "EnsnAired," or "FriendEnergy"?

The design of heat recycling works like this: first the station's ventilation system captures the commuters' body heat, which it uses to warm water in underground tanks. From there, the hot water is pumped to Kungsbrohuset's pipes, covering a third of its fuel needs per year. Kungsbrohuset's design has other sustainable elements as well. The windows, angled to allow in maximum sunlight during the winter, also block the fiercest rays in summer. Fiber optics whisper daylight from the roof into dark stairwells and other nonwindowed spaces, where lazy buildings would need to pay for electricity. In summer, bone-chilling lake water flows through the veins of the building. If you can't cool off regularly by dunking in the lake, at least you can enjoy a sort of dry plunge.

Part of the appeal of heating buildings with body heat is the delicious simplicity of finding a new way to use old technology (just people, pipes, pumps, and water). It's worth noting that the buildings can't be more than two hundred feet apart, or too much heat would be lost in transit. The essential ingredient is a reliable flux of people scuttling to and fro each day to tender the heat, so the design only works in high-traffic areas. Perhaps, on low-volume days, children might be invited to use the space as a gym for high-energy sports.

"Be a joule," the Public Service billboards for EnsnAired heating might urge, noting in fine print: "A person radiates about 350,000 joules of energy per hour, and since 1 watt equals 1 joule per second, one person can effortlessly illumine the darkening world with the energy of a 100-watt lightbulb. A city of 2.25 million people can light 22,500 lamps." Linger with that image for a moment—a multitude of lamps, each one sparkling, but together providing a great cloak of light. Paraphrasing a proverb attributed to Peter Benenson, founder of Amnesty International, it might say: "It's better to become one lightbulb than to curse the darkness."

Widening their vision to embrace neighborhoods, Jernhusen engineers talk of finding a way to capture excess body heat on a scale large enough to warm homes and office buildings in a perpetual cycle of mutual generosity. Heat generated by people at home at night would be piped to office buildings first thing in the morning, and then heat shed in the offices during the day would flow to the residences in the late afternoon. Nature is full of life-giving cycles; why not add this renewable human one?

In this Golden Rule technology of neighbor helping neighbor, we would all share heat from the tiny campfires in our cells—what could be more selfless? Just by walking briskly, or mousing around the shops, you can stoke the heat in someone's chilly kitchen. Possibly a friend's, but not necessarily. I'll warm your apartment today, you'll warm my schoolroom tomorrow. It's effective and homely as gathering together in a cave. Sometimes there's nothing like an old idea revamped.

It's hard not to admire the Swedes' resolve, but it wasn't always this way. During the 1970s Sweden suffered from pollution, dying forests, lack of clean water, and an oil habit exceeding any other in the industrialized world. In the past decade, through the use of wind and solar power, recycling of wastewater throughout eco-suburbs, linking up urban infrastructure in synergistic ways, and imposing stringent building codes, Swedes have axed their oil dependency by a staggering 90 percent, trimmed CO_2 by 9 percent, and reduced sulfur pollution to pre–World War I levels. It was Sweden that in 1968 proposed a U.N. conference to focus on how we're using and depleting the environment, and when it came about, in 1972, Stockholm hosted it. Billed as the first United Nations Conference on the Human Environment, it stressed that *human* and *environment* are no longer separate entities, because we've reached the stage where "man is both creature and moulder of his environment."

In addition to harvesting human warmth with élan, the Swedes excel at coaxing energy for their cities from other renewable sources. Greeting visitors, the world's largest energy storage unit lies beneath Arlanda, Stockholm's airport, where an underground reservoir over a mile long heats and cools the five million square feet of terminals.

On the windswept coast of Sweden, Joakim Byström's company, Absolicon, has developed the world's first solar concentrator that produces electricity and heat at the same time. Made of iron and glass, its shiny tents track the sun like parallel rows of flowers atop Absolicon's roof, fueling its factory. A world away, in a remote region of Chile's Patagonia National Park, twenty of Absolicon's solar collectors fuel the hotels where hikers can overnight in comfort. Lodged atop a hospital in Mohali, India, the company's panels produce heat, electricity, and steam.

In their own sociable way, the small Swedish eco-city of Kalmar and its neighbor towns (a population of nearly a quarter of a million) are making a dramatic change—switching from oil, gas, and electric furnaces to recycled fuel. Heavily forested, the historic port city nestles partly on the Swedish mainland, edging the Baltic Sea, and partly on islands connected by weblike bridges. An important trading city since the eighth century, it combines cobblestone streets with state-of-the-art chic offices and museums. In winter it's hard to tell the shards of ice floating on the sound from the jagged spires of Kalmar Castle, whose reflection mingles with them in the water. With the plentiful forests comes timber, and from it, sawdust and other wood waste, which can be used to create shared *district* heat, in which superheated water is piped through an underground network. Ninety percent of the region's electricity needs are met by hydro-electric, solar, nuclear, and wind power. City-owned cars and buses run on gas made from such pickings as chicken manure, wastewater sludge, household compost, or ethanol. Hybrid cars and trucks patrol the streets, bicycles abound, and low-energy streetlights glow warmly in the dark. Without stinting on warmth or abandoning their cars, the people of Kalmar are proudly drawing 65 percent of their energy from completely renewable sources.

To achieve this, the changeover is happening at every level, in big companies and in small kitchens and living rooms. Many homes and other buildings rely on environmentally friendly district heating. The Soda Cell wood pulp mill, previously known for making oil heaters, switched to renewable furnaces and heat pumps (doubling its sales in the process). Before, the company used to dump its hot wastewater into cooling ponds and release vast clouds of steam into the frosty air. Now it harnesses the steam to drive turbines and the hot wastewater to fill furnace pipes—providing electricity and heat, not only to its own plant, but to twenty thousand homes in a nearby town.

Ideally, every home in the city would have solar panels and electric cars. Kalmar's lofty goal, a community project, is to rid itself of all fossil fuels by 2030. Then, relying on gas, diesel, and oil will be a bygone folly. That's a practical dream with pipes, not a pipe dream, even if it won't happen overnight. "It's important to have small victories," says Bosse Lindholm, who manages Kalmar's sustainability efforts. "It's more important to go in the right direction at a slow speed than in the wrong direction at high speed." Lindholm feels sure the Kalmar model would work elsewhere, "because the challenge isn't technological, it's changing the way people think."

Not far from Kalmar, a dazzling otherworldly mirrored dish perches aloft like a UFO, its landing legs extended. As if to kindle a giant bonfire or torch distant ships, Ripasso Energy has erected the parabolic mirror to catch and magnify the sun, which drives the pistons of a Stirling engine that, as the company's spokesman explains, is "a contraption invented by a Scottish priest in the early 1800s and then further developed by Swedish submarine manufacturer Kockums." The result is a world-record-setting blast of solar energy, the most efficient thus far. Comparing this design to China's Three Gorges Dam, the largest hydroelectric plant in the world, Tore Svensson from Ripasso points out that Three Gorges requires a thousand times more land to generate the same amount of energy. Ripasso's plants produce a hundred thousand gleaming sun-catching mitts a year, which provide as much energy as five nuclear power plants.

I'm spotlighting the Swedes because they're working with such limited raw material, wickedly little sun, and yet they've cooked up all sorts of brilliant designs. The lesson is: you don't need bright sun if you have bright ideas and a culture that promotes them.

Central Station's heat sharing and the eco-hubs in the countryside are just pieces of Sweden's larger sustainability jigsaw puzzle, in which a particularly striking piece is how the country handles its waste. In Sweden, a whopping 99 percent of household refuse is recycled or used to generate energy. Only a dash of it goes into landfills; the rest is scrupulously collected and burned in incinerators with state-of-the-art filters, which generates electricity for a quarter of a million homes and furnishes heat for 20 percent of the country's heating grid. There's just one problem. The Swedes aren't producing enough waste to keep the generators burning. The odd-sounding solution is that Sweden imports eight hundred thousand tons of trash each year from Norway and elsewhere in Europe. Norway pays Sweden to dispose of its trash, and Sweden reaps more electricity and heat. Norway's trash isn't always pristine (whose is), so, not wanting to add pollutants to its shores, Sweden captures toxic chemicals and metals from the ashes and ships those back to fill Norwegian landfills. Germany, the Netherlands, and Denmark also import waste from other countries to keep their incinerators churning out electricity.

We're just beginning to explore the frontiers of harvesting novel sources of fuel. Consider skimming energy from train travel, based on the principle of the pinwheel. As trains pass, a sirocco of hot, jumpy winds follows, spinning up dust devils and chasing newspapers down the platform. They might be blowing from North Africa across the Mediterranean to Southern Europe. *I could do something with that* tugs gently at the mind, in the same spirit that some ancestor, inspired by how an animal track holds water, thought *I could use one of those.* Leveraging the wind, three South Korean designers, Hong Sun Hye, Ryu Chan Hyeon, and Sinhyung Cho, have found a way to use the whoosh of trains to power cities. Their "Wind Tunnel" is a network of underground subway lines that capture the wind roiled up by passing trains and funnel it to turbines and generators embedded in the subway walls. Above ground, there'd be less traffic coughing and guzzling if more commuters relied on trains, and below ground, in the arterial hum of the city, Wind Tunnels would flute electricity to apartments and offices.

China, home to a vast network of high-speed trains, is also tempted by wind-catchers. If the idea fits well there, it may be because pinwheels have figured in Chinese culture and temples for thousands of years as powerful symbols of turning one's luck around by casting out bad fortune and gusting in good. Hidden between the railway ties, wind turbines would funnel electricity into a well for the energy grid, which the trains tap, completing the circle.

Or here's another way to reuse train power: bank it. As I drive around town in my aging Prius, I know that whenever I push on the brakes it banks electricity into its battery. During low-speed driving, it milks the stored power, and it burns gas only at high speed on the highways. This spending and saving balances well, and I rarely buy gas. In Philadelphia, the Southeastern Pennsylvania Transportation Authority joined forces with Viridity Energy to create hybrid subways with the same thrift. Each time a train brakes around a curve or entering a station, it deposits energy into a big battery connected to a shared grid.

Worldwide, in such ways, the outdated idea of travel serving only to carry people from one place to another is gradually melting into the notion of piggybacking and recycling—transportation with bonuses. This pertains to cars and buses, of course, with companies aiming for ever greater mileage on ever less fuel of a preferably renewable sort such as hydrogen or electricity. A new twist on that is the Green Apple concept car, so named because it's designed for use as a taxi in New York City, offering "street hails" in all five boroughs. Not adding to the carbon footprint, it could actually erase part of it. A three-seater shaped like an aerodynamic space helmet, it's powered by turbines that whip in polluted air and purify it before exhaling it back onto the street. A snarky air-scrubber. Remember riding on the vacuum cleaner Mom or Dad propelled? Yes, the air could be called what it is, "recycled waste," but where's the fun in that?

Speaking of fun, some wind-harvesting ideas look like they've sprung from either an aviary or pages of sci-fi. I have several favorites at the moment. One is Windstalk, created by the New York firm Atelier DNA to provide clean energy for Masdar City, Abu Dhabi. A work of "land art," it aims to provide wind energy while fluttering, oscillating, vibrating, and generally behaving "as chaotically as possible," and also being beautiful. The gentle, incantatory winds of an otherworldy oasis infuse the designers' description with irresistible hints of lounging and longing:

Our project starts out as a desire, a whisper, like grasping at straws, clenching water. Our project takes clues from the way the wind sways a field of wheat, or reeds in a marsh. . . . Our project consists of 1203 stalks, 55 meters high, anchored on the ground with concrete bases. . . . The top 50 cm of the poles are lit up by an LED lamp that glows and dims depending on how much the poles are swaying in the wind. When there is no wind . . . the lights go dark . . . When it rains, the rain water slides down the slopes of the bases to collect in the spaces between, concentrating scarce water. Here, plants can grow wild. . . . You can lean on the slopes, lie down, stay awhile and listen to the sound the wind makes as it rushes between the poles. But our project isn't just desire.

Another of my favorites, in contrast, won't sing and dance in the desert like Windstalk. It's more like a sweaty air-cooled wallflower. On the lawn at the Technology University in Delft, I spy what looks for all the world like a time portal, a large sleek steel rectangular window frame hovering in the air, but not quivering in the cool spring breeze. Because it stands outside the Electrical Engineering, Mathematics, and Computer Science building, you'd be forgiven for thinking it resembled a dot-matrix zero. It might be a futuristic sculpture. Yet it's a bladeless, bird-friendly, bat-friendly wind turbine, designed by TU Delft and Wageningen UR, the architecture firm Mecanoo, and a consortium of others as part of a government alternative-energy project. At this stage only a pioneering prototype, it promises a way to reshape wind power into electricity, despite having no moving parts, casting no intermittent shadows, creating no bone-twitching vibrations, and making none of the risen-zombie sounds reported near traditional three-bladed wind turbines coating the flanks of Andalusia or Cape Cod. No wonder passing students stare and instinctively smile, perhaps thinking as I do: *How did you say again that works?*

Fortunately, Dhiradj Djairam and Johan Smit, two Delft professors who helped design the technology, can tell me. Technically, it's a windmill, designed to produce power by milling the wind, a famous part of Dutch tradition—but without moving blades. "Wind energy is converted to electrical energy by letting the wind move charged particles against the direction of an electric field," Djairam explains.

A fluid steel frame encircles horizontal tubes, arranged like venetian blinds, that create tiny electrically charged water droplets. As the droplets are born and rapidly blown away by the wind, they scratch out an electrical current that can flow into a city's energy grid. Round, square, or rectangular, standing alone atop a tall wind-whipped building or in regimental rows along the coast, these

"ewicon windconverters" may become vanishingly familiar the way
TV aerials do, but for a while anyway they'll appear wondrous as
giant bubble-blowing wands. Or as time portals to a future where,
by entangled logic, you remain yourself with all of your individual
quirks and foibles, plus the savvy and ignorance of our age, and yet
sense how future Earthlings live, surrounded by and largely oblivious to a phantasmagoria of techno-wonders that are as commonplace to them as ours seem to us. What sources of renewable energy
will they have mastered? How will they corral the wind and drive
the chariots of the sun?

Imagine Olivine, our future geologist, standing on the shore of a
sea, in the heart of a metropolis or in low orbit, and looking back at
our age. *Those early Anthrops*, she thinks. *How did they live with so much
illness, and so many natural disasters, while polluting all over themselves?
And why did it take them so long to discover*—here you can fill in the
blanks—*the heliocopter, the bladeless wind-thresher, the hydrogen Coupe
de Ville?*

Meanwhile, TU Delft is working on other forms of airborne
windpower, including a "ladder mill," which is really a string of kites
whose blades ride the high winds aloft. "If we move away from the
idea that a turbine ought to have a steel foot," Djairam says, "we can
harvest this wind for our electricity supply."

When it comes to reaping energy from the eye-scalding power-
house of the sun, we've only begun to explore its promise of benefi-
cent fury. Over the millennia, humankind has worshipped the sun,
and with good reason, but these days we rarely pause to marvel at
how it charms our existence. It reaches into the mumbling corners
of our private universe, spurs growth, sheds light on all our episodes
and exploits, transfigures daily life. Its edible rays feed the green
plants on land and sea, which animals graze upon, and we dine upon
in turn, and so it quivers through our blood. Every molecule of our
being, every mote inside us, every atom and eave in the mansion of
the body and the penumbra of the mind was forged in some early
chaos of a sun. It's only in death that our long conversation with

the sun ends. Other elements in our world may trace their origin to lesser luminaries—the gold we mine, for instance, to a sparkling bombardment of asteroids two hundred million years ago. But the sun's breath made all of life possible.

You'd think that would be enough for one species of upright ape, but we rack our sun-smelted brains to find newer ways to capture and enslave the sun to power the rest of our lives. We've been exploiting it throughout the Anthropocene whenever we've burned fuel— really a form of buried sunlight—to warm ourselves and power our empires. The Industrial Revolution always was about solar power. Now we're just skipping the secondhand part and going straight to the wellspring of that fuel. Wood, coal, oil, and gas were only intermediaries after all, and using them was a sign of our immaturity as a species.

Sweden's Ripasso Energy is not the only endeavor beginning to excel at solar power, even if it's not yet as profitable as fossil fuels. It will be, because it *must* be, and soon, if we're to survive all of our masterpieces and conquests. In Nevada, Ivanpah, the world's largest solar thermal facility, already stretches to the horizon in the Mojave Desert. And it should. America receives as much sunshine as light-spangled Spain, the sunniest country in Europe and the world's leader in concentrated solar power. In the coming years, Desertec, a far-reaching $400 billion project, plans to harvest solar energy from Africa's sun-drenched deserts and pipe it to the world. Ample sunlight falls on North Africa each day to power the whole continent, as well as Europe, and Desertec's ultimate goal is to collect enough sunshine in deserts to power the entire planet.

In Germany, solar panels line rooftops like glossy guitar picks, sparkle with pent-up power beside the railways, spangle like beaded frocks on the hillsides, escort cars along the autobahns, stand on stalks and peer up at the sky like sunflowers. They're everywhere one looks, pulsing from inner-city apartments to barns and old abandoned military bases. In the Gut Erlasee Solar Park, where straggling weeds climb between the panels, threatening to shade them, a maintenance crew of grazing sheep dutifully prunes the intruders. In the southern German state of Bavaria, home to 12.5 million people, three solar panels take up residence for every human. While Germany doesn't get an enormous amount of direct sunlight, on one prismatically sunny day in May 2012, it harnessed 22 gigawatts of energy from the sun—as much bottled lightning as twenty nuclear power plants could create, half of all the solar energy being collected around the world that day.

Thanks to shrewd legislation passed in 1991, including financial incentives and widespread support from a citizenry well tutored in the need for renewables, Germany has become a world leader, harvesting wind, water, and sun power for a quarter of its energy needs, with solar providing the lion's share, and German companies spearheading solar technology research and design. The sun's rays may be free, but they're not cheap to use. Solar energy still costs more than fossil fuels or nuclear energy, but prices have fallen 66 percent since 2006, making it obvious that trained sunbeams will soon be as affordable as coal. Meanwhile, solar research is heavily subsidized by the government and also heavily backed by investors. But even without government subsidies, solar energy is flourishing in India and Italy, and China is surfing the solar energy wave with such flair that some German tech companies are being eclipsed by suddenly plummeting prices. Ideally, every home would have solar panels and affordable fully electric cars that could be plugged into the sun, a molten socket that stretches 92,960,000 miles.

Many communities and countries around the world are finding creative new ways to harvest and reuse energy, but most grassroots initiatives aren't covered by the media; even though they may be life-changing locally, to the rest of the world they're invisible. Dayak villagers in Borneo are replacing their diesel generators with hydrogen ones and hydroelectric energy (from streams) to power their lives. In Curitiba, Brazil, once crippled by traffic, 70 percent of commuters now travel by bus, saving twenty-seven million liters of fuel a year and lowering air pollution.

Climate change has become so visible, and wildlife and fresh water so much scarcer, that fewer people are foolish enough to deny the evidence. As we wade into the Anthropocene, we're trying to reinsert ourselves back into the planet's ecosystem and good graces. Unlovely as the word "sustainability" may be, it's sashaying through the media, taking root in schools, and hitting home in all sorts of domiciles, entering the mainstream in both hamlets and megacities. We're undergoing a revolution in thinking that isn't a reaction to the Industrial Revolution, nor is it a back-to-the-land movement of the sort that became popular during the Great Depression and again in the 1970s. We might sometimes resemble startled deer in the head-lights as we face Earth's dwindling resources, yet at the same time we're opening a door to a full-scale sustainability revolution. Our fundamental ideas about house and city have begun evolving into the smarter, greener matrix of our survival.

PART III

IS NATURE "NATURAL" ANYMORE?

IS NATURE
"NATURAL" ANYMORE?

I am writing this in a bay window that floats halfway up an opulent old magnolia tree, which unfolds waxy pink brandy-snifter flowers in the spring, and offers lofts to wrens and chickadees, perches for owls and wing-weary hummingbirds, syrup for yellow-bellied sapsuckers, leafy pounce-ways for squirrels. Its neighbor, a colossal sycamore, hunches dozens of crooked branches to the sky, and catches sunlight in fuzzy leaves the size of bear paws. The deer turn to it for shade, the brown bats for shelter, the goldfinches for edible ornaments. Both trees fork and flow like river systems of sap with many tributaries. They bargain with insects and animals, keep their own time, and possess impulses and know-how I barely understand. Brainless the trees may be, but they have tiers of memory, powerful urges, skills, and faculties. We're all offspring of one crusty planet, but we're so different that we sometimes seem to inhabit alien universes. Even the criminal mind is more explicable than a tree—a quiddity we cannot enter, an essence that does not include us.

Like most other people, I find the magnolia, the sycamore, and the animals part of a wild green spontaneous expanse, where other creatures with other pedigrees are busy pursuing their own cycles and mysterious purposes. In a human-centered world, the otherness of nature is part of its great comfort and allure. For Bill McKibben, "nature's independence *is* its meaning; without it there is nothing but us." Ancient beyond our imaginings, nature offers us a refuge from human affairs, a world free from social labyrinths, romantic tangles, hopes and hurdles. Or so it seems. But is this really true?

My magnolia belongs to a genus more ancient even than bees (beetles pollinated its ancestors), and certainly older than humankind. The story fossils tell is that a magnolia can trace its roots back a hundred million years. Its family has survived ice ages, the uprising of mountains, and continental drift. Its lineage may be older than the Finger Lakes hills. Yet it didn't begin its life in my chilly yard. Aztec admirers named it Eloxochitl, the tree with green-husked flowers. Spanish explorers in the New World, enchanted by the large waxy petals that blushed like a maiden's cheeks, ferried magnolia roots home along with the new luxuries of chocolate, vanilla, and brilliant cochineal-red dye. By the 1730s magnolias adorned many European gardens and were hybridized with other species, evolving into the robust ornamental magnolias that ultimately voyaged back across the Atlantic to grace southern homesteads, where they were usually planted in the front yard. In time horticulturalists sold their novelty to northern nurserymen, one of whom sold it to the original owner of my property, an entomologist, who no doubt watered, fed, and tended it lovingly.

This stately old magnolia is so bound up with human schemes and follies that it's not exactly "wild" but rather part of our manmade world. It's more akin to a domestic animal that lives in partnership with humans, providing beauty and a remembrance of the wilderness. The same is true of the sycamore. Although I live atop a hill, sycamores usually grow on the margins of rivers, or in wetlands, thriving on green banks between a field and a stream. Opossums, wood ducks, herons, and raccoons nest in a sycamore's many cavities and branches. Native Americans sometimes used the entire trunk of one tree to carve a dugout canoe. It's covered in apple-shaped fruits, each one a tiny Sputnik tufted with brown hairs and full of seeds. But my sycamore is really a disease-resistant hybrid of an American and an Oriental sycamore. So it, too, is a traveler, or at least its genes are.

As for the wild birds, I feed hummingbirds sugar water and put out seed for the dark-eyed juncos, nuthatches, and finches. Many of the crows wear armbands, as if in perpetual protest. They're being studied by local ornithologists, and each band bears numbers, a favorite human logo. The tags are applied with care, and I don't think they hamper the birds. I sometimes see a crow preening its tag into place as if it were another feather. But, like the trees, the birds don't live detached, independent lives. Humans have meddled with their whereabouts, numbers, health, and gene pool.

In contrast to life indoors, I regard this landscape full of birds, trees, and animals as "nature." From that perspective, the telephone poles and fence on the property line, the TV cable and metal mailboxes, the asphalt street and grumbling cars and distant arpeggios of downshifting trucks, all belong to the crafted world of humans, an artificial paradise filled with ceaseless blessings and hardships.

The myth of our sprawly, paved-over cities and towns is that we've driven native animals out and stolen their habitat. Not entirely true. We may drain the marshes, level forests, and replace meadows with malls, exiling some animals. But, because we also need nature, we create a new ecology that happens to be very hospitable to wild animals. For a few species, it's more inviting than wilderness. Our buildings offer cubbyholes and crevices for animals to nest in. We install ponds, lawns, groves of edible trees. We leave garbage on the curb and design flower beds that are well watered and well fed, serving a smorgasbord of delicacies easily within a deer's reach. In the process we keep fashioning new niches, most often without meaning to.

Anthropocene cities have created pools of a limited number of species, the ones that coexist well with humans—mainly deer, rats, cats, birds, foxes, skunks, raccoons, houseflies, sparrows, mice, and monkeys. One finds such city species wherever animals are forced to live in our shadow, feeding on our leavings, and joining the fossil record beside our steel and plastic. But we're restyling their evolution, because urban animals (including humans) vary their habits and psychology to adjust to city life. Animals living in parks and zoos also adapt to our natural biorhythms and landscapes.

As more birds harbor in the cities, they find plenty to eat, but their biological clocks skip ahead. When Barbara Helm, a University of Glasgow ornithologist, compared blackbirds in Munich with their country cousins, she found that city birds start their workdays earlier and their biological clocks tick faster. Just like their human counterparts, they adopt a faster pace, work longer hours, and rest and sleep less in cities where upward-showering light washes out the stars and our handmade constellations cluster near the ground. Urban males also molt sooner and reach sexual maturity faster. In contrast, country blackbirds begin their day traditionally, at sunrise, don't rush, and sleep longer.

"Our work shows for the first time," Helm concluded, "that when sharing human habitats, a wild animal species has a different internal clock."

Her colleague on the study, Davide Dominoni of the Max Planck Institute for Ornithology, added that for city songbirds, "early risers may have an advantage in finding a mate and thus a greater chance of successfully producing offspring and passing along their chronotype—the time of day their body functions are active—to the next generation. Other research has shown that chronotypes are highly heritable, so the process of natural selection could mean that city birds are evolving to favor early risers."

Tinkering with evolution, we subject our pets and plants (as well as the wild animals who live near us) to our manufactured schedules of light and dark, sleep and waking, toileting, exercise, and feeding. Seasonal time has given way to a chronicity which has its own intricately satisfying beauty and a certainty one rarely finds in nature. We've not only rigged clocks to slice our days into tiny even segments, and lit up the night with noble gases trapped inside glass, rewiring our own circadian rhythms in the process, we're also resetting the rhythms of the planet's other life forms.

In Aesop's fable "The Town Mouse and the Country Mouse," two cousins exchange visits, during which the city mouse turns down his nose at humble country fare, and the country mouse discovers that city life, while richer, is unbearably dangerous. *I'd rather gnaw a bean than be gnawed by continual fear,* he wisely opines. But thanks to us, today's city mice are growing big brains to outwit the ambient dangers. Not just mice. According to researchers at the Bell Museum of Natural History at the University of Minnesota, we've caused at least ten urban species—including voles, bats, shrews, and gophers—to grow brains that are 6 percent larger than those of their country cousins. Heavens, smarter rats! That's a scary thought. As we felled and planted over their woods and meadows, only the cleverest animals survived, by tailoring their diet and behavior to the human-dominated landscape. Those who did passed big-brained genes on to resourceful offspring. And they were the lucky ones. Not all plants and animals can evade us or evolve; only the most flexible endure.

To cope with urban life, some animals have even begun redesigning their bodies at a pace fast enough for biologists to track. On a flat, horizonless Nebraskan highway, Charles Brown will often pull over to inspect a fresh piece of roadkill, provided it's a cliff swallow. Chestnut-brown-throated, with white forehead, pale breast, and long pointy wings, cliff swallows favor cliffs, their ancestral roosts. I've enjoyed watching aerobatic crowds of them barnstorming the cliffs of Big Sur, where their calls—banshees quarreling in high, squeaky twitters—mix with crashing surf.

But cliff swallows do need cliffs. These days, faced with city sprawl, they're plastering their gourd-shaped mud nests onto buildings, beneath highway overpasses, and tucked into railway trusses and trestles, building up colonies of thousands on our concrete cliffs.

A behavioral ecologist at the University of Tulsa, Brown has been observing their gregarious social life for thirty years, traveling from colony to colony, and often passing birds killed in the maelstrom of traffic. He'll stop and check for a leg band and perhaps collect the bird for research.

"Over time," he says, "we began to notice that we were seeing fewer dead birds on the roads."

A bigger surprise was the length of their wings. The roadkill birds had longer wings than the swallows he'd caught in mist nets. These two changes—fewer birds dying in accidents, and a difference in the wing length of dead versus living birds—led him to a startling conclusion. To cross the road safely, cliff swallows had to weave and dodge at speed, favoring those with the short wings of dogfighting jets. The unlucky swallows with long wings more suited to pastoral life died in accidents, leaving the short-winged swallows to breed and become dominant. All in just a few decades.

"Longer-winged swallows sitting on a roadside probably can't take off as quickly, or gain altitude as quickly, as shorter-winged birds, and thus the former are more likely to collide with an oncoming vehicle," Brown suggests. "These animals can adapt very rapidly to these urban environments."

How should we regard the blackbirds, cliff swallows, and other animals that are evolving in such a snap because of our technology? Will they become new species? Or are they just new citizens of our age?

What makes nature natural? It's a quintessentially Anthropocene question. Nature thrived long before cities did, long before we coated the Earth with an immensity of humans. Wild animals live among us. Our toil and our machines are entwined in their fate. Even our densest city is a permeable space, although we try hard to live a world apart. We decide the limits of the wild and where a city begins and ends. Suburban sprawl has replaced the overgrown buffers we used to have, transitional land between the two worlds. Now wild and urban animals encounter one another daily.

We cherish a strong sense of place, rich with memories. But other animals abide by a sense of place, too. Banding studies show that ruby-throated hummingbirds travel the same route every year, zig-zagging to their favorite yard. A familiar pair of mallards comes to canoodle behind my house every spring. Countless other critters return to a special mating or nesting spot, and will continue trying, even if we fragment their world on a grand scale by installing the materials, plants, and animals we prefer. When we claim a patch of real estate, scent-marking it with our stuff, and purging it of wild animals, we presume the animals will bow out graciously. As sensitive tyrants, it rattles us when they don't and try to resettle their once-cherished digs.

Citywise animals are mainly invisible to us, hunting at night or creeping in shadows, and if we do encounter them, they surprise us by being out of place. We forget that the animal kingdom is a circle of neighbors who often drop by unannounced. Even if the previous residents have skedaddled, or rerigged their schedules, new species may begin showing up like furtive relatives from who knows where. By the time you realize they're not just visiting, they've shot down roots, claimed a little fiefdom, disturbed some of your neighbors, and added a tiny codicil to daily life. Not always a welcome one.

Before the 1990s, no one saw coyotes on the streets of Chicago. Now the city offers refuge to two thousand, which prefer parks, cemeteries, and ponds and generally flee from people. But some have been tracked crossing more than a hundred roads a day and moving into residential neighborhoods. Moose regularly pay house calls in Alaska, stomping into yards and onto porches, looking for grub. Giant antlers and all, they can leap chain-link fences. On many a golf course in Florida, alligators create an extra water hazard, and lakeside settlers know to keep their Chihuahuas indoors. Mountain lions forage in Montana cities; cougars stalk joggers in California; elk stroll through housing tracts in Colorado. When one Jackson-ville woman lifted up her toilet seat, a water moccasin leapt out and bit her; another woman, in Brooklyn this time, found a seven-foot-long python in her toilet. Leopards prowl the streets of New Delhi by night. In the Royal Botanic Gardens, in Melbourne, Australia, endangered gray-headed flying foxes built a colony of thirty thousand bats, drawn to the garden by all the cultivated native plants: eighty-seven healthy tree species with the fruit they preferred, a year-round oasis. Why risk the outback? And, maybe strangest of all, prairie dogs, those ground-dwellers of the open range, have begun digging their towns in our cities.

However, as it's beginning to dawn on us that there's no sharp line between the untamed and the built up, more people are trying to help wayward creatures find their way through our mazes.

When a lost eight-month-old coyote strayed into downtown Seattle, he became confused by the streets and buildings and grew frightened and disoriented. Dashing for what must have looked like a dark haven, he ran through the open door of the Federal Building, skidding on polished floors and around narrow hallways, bumping into glass, walls, and people in a panic. Then he spotted a cave to hide in—an open elevator—and darted inside, and the doors closed. For three hours, the poor creature paced that metal box until people from the state fish and wildlife department trapped him and set him loose outside of town.

It's surprising how disruptive even a slow, lowly terrapin can be. One June day recently, more than 150 diamondback turtles scuttled across Runway 4 at JFK, delaying landings, halting takeoffs, foiling air traffic controllers, crippling timetables, and snarling air traffic for over three hours. Cold-blooded reptiles they may be, but also ardent and single-minded. Never mess with a female ready to give birth.

Graced by beautiful rings and ridges on their shells, diamondbacks look like a field of galaxies on the move. We think of the shell as a lifeless kind of armor, but it's actually attached to their nervous system, not just a bulwark but an integral part of their inner world. They inhabit neither freshwater nor sea but the brackish slurry of coastal marshes. Mating in the spring, they need to lay their eggs on land, so in June and July they migrate to the sandy dunes of Jamaica Bay. The shortest route leads straight across the busy tarmac.

Don't the plucky turtles notice our jets? Probably not. Even with polka-dot necks stretched out, diamondbacks don't peer up very high. And unlike, say, lions, they don't have eyes that dart after fast-moving prey. Ploddingly slow, they abide by seasonal time, so the jets probably blur into background—more of a blowy weather system than a threat. But planes generate a lot of heat, and the turtles surely find the crossing stressful. Not to mention the roundup. After a little light banter between pilots and air traffic controllers, Port Authority crews descended, scooping turtles into pickup trucks and ferrying them to a nearby beach.

"We ceded to Mother Nature," said Ron Marsico, a Port Authority spokesman. "We built on the area where they were nesting for generations, so we feel incumbent to help them along the way."

Mounted on the shoreline of Jamaica Bay and a federally protected park, indeed almost surrounded by water, JFK occupies land where wildlife abounds, and it's no surprise that planes have collided with gulls, hawks, swans, geese, osprey, and even milky-winged snowy owls (an influx from the Arctic). Or that every summer there's another turtle stampede, sometimes creating lengthy delays. As a private pilot, I remember well how airports used to treat animal "hazards"—at gunpoint. It's heartening these days to find other solutions, from relocation to relandscaping, with canny coexistence the preferred option.

In my town, we're blessed by lots of wild animal visitors, from star-nosed moles and eagles to otters, wild turkeys, foxes, and skunks. White-tailed deer are so numerous that they qualify as residents. Last week I was shocked to see a coyote toe stealthily up to the bird feeder outside my kitchen window, below which sat a plump seed-gobbling rabbit. When I opened the window to address the coyote, he turned tail and trotted into the tall grasses lining the driveway. Yesterday evening I caught sight of him once more, this time as a streak of yellow dots and dashes weaving through the bushes in my backyard. It took a moment for my brain to decode the pattern, and another moment to start worrying about the two baby rabbits eating clover on the lawn.

On a rainy morning so gray a dappled mare could get lost in it, my village held a public hearing to decide the fate of our local deer. Over a hundred residents spoke out against the proposed amendment to the firearms law, which would invite wildlife exterminators to bait and shoot the deer as long as they were at least five hundred feet from houses, schools, and yards. Lured with corn, the deer would be killed by high-powered bows and rifles. Because ricocheting bullets and arrows are possible, the village plans to take out liability insurance in the multimillions. If this sounds like a dangerous and extreme solution to the deer problem, you'll understand the passion of the protesters.

Homeowners defended shooting the deer, which they regard as vermin. For them, it's either the deer or the landscaping. Several gardeners conceded that deer had eaten many of their plants, but argued in favor of deer fences, not gunfire. One man grew tearful as he implored the board to live in harmony with nature. A psychologist accused the board of "groupthink," in which deer have become a new demonized minority. Mothers worried over the safety of children walking home from school or playing outside amid stray bullets—and also over the psychological damage of witnessing the death of wounded deer.

One little girl asked her mother: "If they shoot all the deer, how will Santa deliver the presents?"

Another mother said that in her child's elementary school, peaceful arbitration was being taught. She asked: "How can I begin to explain the hypocrisy of grown-ups solving their deer problem by hiring killers to gun down the deer?"

"There are so many deer fences—it's like living in a war zone!" a kill-the-deer man cried. To which a save-the-deer man replied: "And you think snipers firing bullets around the village for the next ten years will be *less* like a war zone?"

Most protesters pleaded with the board to give fences and ster-
ilization a good chance. Others argued that the board was legally
bound to follow majority rule and should start shooting. Some
debunked long-held myths about deer and Lyme disease (the white-
footed mouse carries the agent, and killing the deer won't banish
the Lyme tick, which feeds on twenty-seven species of mammals,
including cats and dogs). Or the idea that deer cause the most traffic
accidents (speeding and alcohol do). Or that birth control methods
fail (immuno-contraception has worked in national parks). Contra-
ception is expensive, but so is hiring sharpshooters every year and
paying for liability insurance.

What struck me as some kill-the-deer people spoke was the tone
of dread and loathing, a panic about being invaded by wildness
and roughly overtaken by the chaotic forces of nature. It's as if we
weren't talking about the deer at all, but about what Freud called the
Id, that wild demon of the psyche we keep just barely in check, and
which otherwise would be slobbering, rutting, and killing all the
time. What if its sheer feral exuberance took charge? Soon, neighbors' yards would teem with tall gangs of unruly weeds. Or they
might stop raking the leaves, and then clots of color would smother
everyone's lawn. Four-legged predators inspire the most panic, but
if wild turkeys and deer can find their way into suburbia, can fiercer
animals be far behind, ones with fangs and teeth, whose red eyes
pierce the night?

Yet, at the same time, something deep inside us remembers being
accompanied by animals. There was a time not very long ago when
cows, goats, horses, and other animals slept indoors beside us, or at
least shared the same roof. In some parts of the world, they still do.
But most humans have pitched their plaster-walled tents in cities and
suburbs, crowding out animals, especially wild animals, and pushing them farther and farther away, to the perimeters of daily life.

In the mists of the mind, we've lost our time-honed knack for
coexisting with other creatures. We erect walls to keep nature out
and take pride in scrubbing dirt and dust from our homes. Then

we adorn our houses with bouquets of flowers, and scent absolutely everything that touches our lives. We seat windows in our walls, install seasons (air-conditioning and heat), and fasten at least one noonday sun in every room to shower us with light. Confusing, isn't it?

Even indoors, we surround ourselves with pet companions who help bridge the apparent no-man's-land between us and nature, between our ape-hood and civilization. A dog on a leash is not really tamed by its owner. It's a two-way tether. The owner also extends himself through the leash to that part of his personality which is pure dog, the part that just wants to eat, sleep, bark, mate, and wet the ground in joy. We've all felt it.

Nature is dynamic and haphazard, and so are we—not a serene combo. Maybe it's one that's best described in paradoxes such as *organized chaos*, but we're not beings who feel comfortable with paradox. Paradox tugs the brain in opposite directions, confounds our quest for simple truths, and throws a monkey wrench into the delights of habit. Faced with paradox, our brain automatically slaves to solve or squash it. And so here we find ourselves, disorderly beings, blessed or cursed with order-craving minds, in a disorderly universe we're fully capable of bringing increased order to—but not absolute order, and not forever.

I SOMETIMES WONDER what Budi would make of our metropolitan jungles. Just like city monkeys the world over, leaping across rooftops, shimmying down drainpipes, nesting at night in the canopy of iron fire escapes, Budi would adapt to the hard surfaces of city life. On school playgrounds, children might see him using the monkey bars and jungle gym with a simian ease they only dream of. He'd find fruit to steal on many corners, densely treed parks where he'd mingle with the other species of great apes, many his own size and mental age, though physically much weaker and easily hurt. Some of the same urban animals that scare us would scare him: bears, coyotes, mountain lions, and such. Would he regard the city as another natural landscape, with blockish mountains, vast herds of humans, and their many watering holes and bazaars? Most likely he would. He'd not only adapt, he'd change his behaviors to suit the new realm, just as so many other urban animals (including us) have been doing with surprising success. It seems obvious that a city, or a cage in a zoo, is not what we mean by a "natural" environment, but in the Anthropocene, it can be hard to say what is.

IF WE DON'T want even more animals living with concrete sidewalks and feeding off human garbage, we must intervene. At this point, preserving the wild is not just a matter of hands-off, as traditional conservation decrees, but also the hands-on of creating other kinds of habitats, such as wildlife corridors. In my mind's eye, I see flashes of the tiny green rainforest on Brazil's Atlantic coast, an Amazon-like realm where the highest concentration of endangered birds in the Americas and the remaining golden lion tamarins cavort in small pristine Edens atop mountains riven by highways and towns. A dozen years ago, when I traveled there as part of the National Zoo's Golden Lion Tamarin Conservation Program, another team was busy building a wildlife corridor, Fazenda Dourada, to link up the mountaintops and extend the birds' and tamarins' range. Not far away, snaking from Argentina all the way up into Texas, the Jaguar Corridor has gifted the scarce, almost mythic spotted cats with space to roam. It only seems fitting that, having rent the fabric of the wild, we at least stitch some green sleeves back together so that animals can rejoin their kin and migrate along ancestral routes. Around the globe countries have been avidly building these links, prompted by a fruitful mix of compassion and self-interest. The United States has some lengthy wildlife corridors, such as the Appalachian Trail, a thousand-foot-wide greenway running two thousand miles along ridgelines from Georgia to Mt. Katahdin in Maine. In India, the Siju-Rewak Corridor protects 20 percent of the country's elephants from collisions with human civilization and its toys. Kenya has created Africa's first elephant *underpass*, a tall tunnel beneath a traffic-snarled major road, which gives two elephant populations long divided by human dwellings a chance to migrate, mingle, find mates, and avoid terrified humans or terrifying traffic.

In Europe, the Green Belt Corridor will soon allow wildlife to ramble all the way from the tip of Norway through Germany, Austria, Romania, and Greece, deep into Spain, following ancient trails while searching for food, mild weather, and safe birthing grounds. Linking twenty-four countries and winding past forty national parks, it spans nearly 8,000 miles, some of it following the old historical line of the Iron Curtain, the 870-mile-long chain of fences and guard towers that once clawed the length of Germany, separating the East from the West. After reunification, what was once "no man's land" lingered as a lifeless scar, until conservationists began reshaping it into a winding nature corridor. Transformed again by human hands, the blue-ribbon path now proclaims tolerance, not repression. The refuge naturally includes many different habitats, from sand dunes and salt marshes to forests and meadows. Ditches that kept vehicles from crossing are being crisscrossed by endangered European otters. Eurasian cranes, black storks, moor frogs, white-tailed eagles, and other stateless species are mingling safely.

In France, China, Canada, and other countries, more corridors are offering tunnels, underpasses, viaducts, and bridges to help wildlife cross their native range, while protecting them from vehicles and us from them. In the Netherlands, six hundred such over- and underpasses allow roe deer, wild boar, European badgers, and their kind to navigate around everything from railway lines to sports complexes. All this adds to a renewed sense of kinship: animals trotting, shuffling, climbing, and winging safely among us, visibly a part of life's seamless web. Like any other close relationship, living with wildlife requires compassion, compromise, and seeking solutions that will benefit all. If peaceful coexistence were easy, there would be no divorce or political strife, only households and empires of domestic tranquillity.

Like many of my neighbors, I fence in the deer's favorites: roses, rhododendrons, day lilies, hostas. In the front yard, I plant beauties the deer reject—iris, peony, cosmos, allium, false indigo, foxglove, monkshood, bee balm, bleeding hearts, sage, daffodil, veronica, poppy, dianthus, and many more—though they still find a lot to munch on. Instead of fencing in the whole property, I've left a corridor for the deer, foxes, coyotes, and other critters alongside a creek that ultimately winds north to Sapsucker Woods.

I enjoy sharing the neighborhood with so much wildlife, a kinship that greatly enriches my life. I'd rather the groundhogs didn't burrow under my study, and the raccoons didn't play chopsticks on the bathroom skylight and stare down with bandit eyes—but I haven't evicted them. I relish the swoop of brown bats at sunset, elegant and enchanting little creatures that eat hundreds of insects every night. Sex-crazed frogs and toads party in the backyard, making a ruckus that can drown out TV or movie watching, but I find their ballyhoo a hilarious part of summer's jug band music. Plying the water below them and adding to the fiendish din are water boatmen, dark copper insects with olive stomachs who swim on their backs, paddling with two oarlike legs, while carrying a silvery bubble of oxygen to breathe as if they were early argonauts. Though small ($\frac{1}{4}$"–$\frac{1}{2}$"), they're adjudged the loudest animals on Earth relative to body size. During sultry summer nights, their singing penises (scrubbed fast over the stomach, washboard-style) can reach 99.2 decibels. That's louder than standing near a freight train, louder than sitting in the first row of a concert hall during a thunderous symphony, even if the water muffles some of their clamor. I'm impressed by the platoon of male spotted newts doing he-man push-ups on the driveway and atop the fence, hoping to make females swoon. I'm delighted when a flicker beats heavy metal tunes on the stop sign—or even if he repeatedly rings my doorbell, as happened one summer. I enjoy spotting red-crested pileated woodpeckers, big as Cheshire cats, whacking the stuffing out of trees. Delving squirrels mean I have to plant bulbs under chicken wire, but I'm amused by their antics. I'm a bit sad I don't have inquisitive black bears to contend with. Deer are the largest animals to pay house calls, and like the dogfighting hummingbirds, tree-climbing chipmunks, and rabbits engaging in an odd tournament of hopping jousts, they arrive unbidden but are welcome emissaries from the natural world.

Each year, I line up behind a dozen cars on a busy highway as a caravan of Canada goose chicks waddles across in a single line between guardian geese, apparently unfazed by motorized honking and the occasional impatient driver. Most people, like me, sit quietly and smile. Like the turtles at JFK, they remind us that, even with egos of steel and concrete plans, we're easily humbled by nature in the shape of snowflakes, goslings, or turtles—all able to stop traffic. They also remind us how conflicted we really are about nature.

IS NATURE "NATURAL" anymore? Of course. But it's no longer indisputably other. The earth scientist Erle Ellis has invented the term "anthrome" to refer to the "hybrid human-natural systems that now dominate Earth's surface." From our small vest-pocket gardens to our giant wilderness areas and parks, nature now reflects our preferences, and one of our most cherished ideas about nature is that nature should be human-free. So we have evicted the indigenous peoples from lands we wished to designate national parks, from the United States' Yellowstone and Grand Canyon to Cameroon's Korup National Park and Tanzania's Serengeti, even though tribes may have lived there for ages, and coexisted to an inspiring degree with the environment.

For Europeans, the word "wilderness" used to mean a wild, barren, chaotic place full of plight and mischief, where it was simple to lose one's bearings or mind. It's easy to forget how ugly nature often seemed to people before Romanticism reexplored the ruggedness of natural beauty. Early-nineteenth-century writers found wil-

dernesses grotesque—not just dangerous and obstructive and rife
with bloodthirsty animals but actually a vision of evil. Now the idea
of wilderness is just the opposite: a sanctuary, an emblem of serenity,
a view of innocence.

Nature is always mutating, on a large and small scale—the lavish
suns of summer, the dragonfly's seasonal demise. Those regular turnovers can become humble as old clothes, nothing to raise
a ripple of awareness, let alone concern, and too rarely a sensory
cascade. The romance with nature—childhood—gives way to the
companionship stage, a time of purposeful beguiling, when it takes
more to capture your attention. But a nonmigration of geese, a
neverthriving of crops, a carillon of snowdrops blooming a month
too early, ripe berries way out of season, a bay full of lobsters ankling
off to the north, the weird absence of winter—these give one pause.
Our newest idea of nature is one of vulnerability, a vast, sprawling,
interlaced organism growing weaker.

At precisely the moment we're achieving unprecedented feats and
ruling the planet on a grand scale, we're discovering that our future
as a species may suffer as a result. Nature isn't separate from us, and
part of our salvation as a species depends on respecting, if not rejoic-
ing in, that simple companionable truth.

THE SLOW-MOTION INVADERS

Named P-52, as if she were a bomber or a precious fragment of papyrus, the Burmese python recently found in the Everglades weighed 165 pounds and stretched 17 feet in length, setting a local record (not a world record—that's held by a 403-pound, 27-foot-long python residing in Illinois). A tan beauty, with black splotches that resemble jigsaw puzzle pieces, dry satiny skin, and a body like a firm eraser, P-52 had a pyramidal head, a brain surging with raw instinct, tiny black Sen-Sen eyes, and a mind like a dial tone. In her heyday, she could squeeze the life out of an alligator or a panther. And she was pregnant.

Standing shoulder to shoulder at the dissecting table, amazed University of Florida scientists uncovered eighty-seven eggs in her womb. Not all the hatchlings would have survived. But with such fecundity it's easy to understand the flourishing of pythons throughout the southern region of the Everglades—slipping through the sawgrass, sibilant as sassafras, slanting up to their prey, and then—slam!—seizing hold with back-curving teeth, crushing and slowly swallowing every morsel.

No one knows precisely how many pythons inhabit South Florida, but reliable estimates run to thirty thousand or more. Over the last ten years, snake wranglers removed 1,825 pythons from as far north as Lake Okeechobee and as far south as the Florida Keys. In the picturesque, if amusingly named, Shark Valley (no sharks, a valley only a foot deep) in the heart of the Everglades, visitors may glimpse a python plying the river of grass, or even wrinkling across the road. Pythons will also be busy hunting, sun-swilling on the canal levees, mating (in spring), coiling around their eggs and trembling their muscles to incubate them, occasionally wrestling with alligators, and absorbing warmth from still-toasty asphalt roads at night.

Alas, they've vanquished nearly all the foxes, raccoons, rabbits, opossums, bobcats, and white-tailed deer in the park; also the three-foot-tall statuesque white wood storks. A survey conducted between 2003 and 2011, and published in the *Proceedings of the National Academy of Sciences,* reported that raccoons had declined 99.3 percent, opossums 98.9 percent, and bobcats 87.5 percent. Marsh rabbits, cottontails, and foxes completely disappeared. Last year, one python was found digesting a whole 76-pound deer.

Where did all the pythons—native to India, Sri Lanka, and Indonesia—come from? Some were wayward pets or hitchhikers in delivery trucks. Others escaped from ponds overflowing in heavy rains, from pet stores during hurricanes, or from international food markets. They hid in foreign packing materials for plants, fruits, and vegetables, or clung to boat hulls or propeller blades. Some may have freeloaded in the ballast of large ships, which take on water and who-knows-what aquatic species in a foreign port, and release alien life forms when they reach their destination. Others sneak a ride on board globe-trotting pleasure or military planes.

Many invasive life forms arrive legally, as desirable crops or companion animals that help to define us or just strike our fancy. Burmese pythons have become popular pets in the United States, credited with a pleasant personality, as snakes go. They're sometimes bred as stunning yellow-and-white mosaics—like the one Britney Spears slipped around her shoulders and slither-danced with at the MTV Video Music Awards in 2001. But many python owners, chastened by the twenty-year commitment, or alarmed by how quickly the reins of power can shift as the snake grows, turn them loose in the Everglades, assuming it will offer an Edenic home. It does. Undaunted by anything smaller than a mature alligator, they eat everything with a pulse, ravaging the whole ecosystem. Native species haven't yet evolved to resist or compete with them, the strongest toughs around.

OF COURSE, MOST of us humans are transplants, too, perpetually bustling between cities and taking familiar plants and animals with us—by accident or by design—without worrying much over the mischief we may be unloosing. We are like witches, leaning over the cauldron of the planet, stirring its creatures round and round, unsure about our new familiars—not wildcats, but pythons?—and waiting to see what on Earth may bubble up next.

Accidental hobos, exotic species travel with us everywhere. A list of known invasive species would fill pages, and their handiwork volumes. Because, like the Burmese python, they can wreak havoc with an ecosystem, we scorn them as marauders, as if it were their fault. But most often we're the ones relocating the planet's life forms.

Invasive species may carry hobos of their own, contagious ones we're not immune to. When a San Francisco woman's pet boa, Larry, fell ill recently, scientists studied the genome of boas and to their shock discovered a genetic mishmash of arenaviruses, which spawn such human nightmares as Ebola, aseptic meningitis, and hemorrhagic fever. It's entirely possible, they surmise, that Ebola began in snakes and spread to humans. Or that, somewhere along the evolutionary road, snakes became vulnerable to Ebola, just as we did. Now we know that reptiles can harbor some of the world's deadliest human viruses, yet we still ferry them from one locale to another.

Pythons aren't the only brawny Floridian invaders. In Cape Coral, monitor lizards—which can reach six feet long—threaten the protected, and altogether winsome, burrowing owl. Gambian pouched rats are overrunning Grassy Key. Cuban tree frogs devour smaller native frogs. Giant African snails dine on five hundred different plants. Jumbo green iguanas are driving the Miami blue butterfly toward extinction. And monk parakeets flock across the Florida skies, flat-nosed as aging prizefighters, making otherworldly shrieks that sound like people prying the lids off cans of motor oil. Unfortunately, their large colonial nests can damage residential trees and electrical power lines, and not everyone is a devotee of squawks, so they're regarded as a nuisance. Florida boasts more invasive species than anywhere else on Earth, from wild boars and Jamaican fruit bats to squirrel and vervet monkeys, nine-banded armadillos, and prairie dogs.

The same thing can happen in freshwater, and the Finger Lakes now teem with zebra mussels (Russian natives) that clog boat engines and water intake pipes and weigh down buoys. In Tampa Bay, green mussels (New Zealand natives) are smothering the local oyster reefs. Asian carp are turning the Great Lakes into their private dining room. With great relish, the rainbow-sheened Japanese beetles are snipping rose leaves into doilies. Although the Nile perch was introduced into Lake Victoria to appear on the tables of locals, no one counted on its predatory gusto, and it's decadently feasting on a hundred species of native fish.

We've been wantonly shuffling life forms for tens of thousands of years. Migrating bands of *Homo sapiens* carried plants, animals, and parasites with them on their travels, and ancient texts often speak of importing exotic delicacies and species from foreign lands. Vagabond species travel in our luggage, cuffs, and cars— shadowing us around the block and around the world. During the seventeenth- and eighteenth-century voyages of exploration, along with ideas and goods we spread vermin and disease. We colonized every continent and all but the most extreme ecosystems, reconfiguring them at speed. Just breezing through our lives—hiking across a meadow, commuting to work, flying or sailing overseas—we keep rearranging nature like a suite of living room furniture.

So invasive species have been running riot for ages, some a plague and a nuisance, others a delight. We've transplanted a great many plants and animals on purpose, for their beauty, novelty, taste, or usefulness—from starlings and poison ivy (a nonallergic European found it pretty and took it home with him) to exotic reptiles and azaleas. Charmed by the climate and organisms at their new locale, they've taken hold, sometimes fiercely (as is the case with eucalyptus, bamboo, and Indian mongooses), to the distress of local species and human residents. People love their English ivy, Norway maple, bullfrogs, Japanese honeysuckle, oxeye daisies, St. John's wort, dog roses, Scots pine, etc. In contrast, such alien invaders as African bees, tiger mosquitoes, fire ants, water lettuce, burdock, lampreys, loosestrife, bamboo, kudzu vine, and dandelions (which apparently accompanied pilgrims on the *Mayflower*) are scorned, cursed, and uprooted.

We insist that invasive species don't belong in wilderness, but native ones do—even if they've died out. Inspired by that notion, we've reintroduced wolves into Yellowstone, moose into Michigan, European lynxes into Switzerland, musk oxen into Alaska, Przewalski's horses into Mongolia and the Netherlands, red kites and golden eagles into Ireland, cheetahs into India, black-footed ferrets into Canada, brown bears into the Alps, reindeer into Scotland, northern goshawks into England, Bornean orangutans into Indonesia, condors into California, giant anteaters into Argentina, Arabian oryx into Oman, peregrine falcons into Norway, Germany, Sweden, and Poland—and a great many more. At the same time we're destroying some ecosystems, we're busy recreating many others.

People may talk about rebalancing an ecosystem, but there is no perfect "balance of nature," no strategy that will guarantee perpetual harmony and freedom from change. Nature is a never-ending conga line of bold moves and corrections. Hence the continuing debate about whether or not the Everglades should be python-free or allowed to evolve into whatever comes next. Ever since the 1920s, we've been transmogrifying Florida swamps into houses. So the real question is what sort of pocket wilderness we prefer.

I'm of two minds about this case. On the one hand, I don't want to disturb the dynamic well of nature. Habitats keep evolving new pageants of species, and we shouldn't interfere. Yet I also sympathize with those who argue that we should capture the pythons in the Everglades and allow the ecosystem to return to its admittedly idealized state, where foxes, rabbits, deer, and a host of other vanishing life forms may flourish. We're losing biodiversity globally at an alarming rate, and we need a cornucopia of different plants and animals, for the planet's health and our own. By introducing just one predator into a beloved habitat, we've doomed a shockingly large segment of species and all those that depend on them.

The tug-of-war we secretly feel between our animal and human natures is part of what makes us endearingly compassionate, and mighty strange primates. Unlike other animals, we care deeply about scores of life forms with whom we share the planet, even though they're not family members, not even species members, for that matter, not possessions, and not personal friends. We care abstractly about whole populations we may not have seen firsthand, determined to help fellow creatures survive. We feel a powerfully mingled kinship.

Whatever interventions or restorations we might plan, our unplanned intervention, in the form of climate change, is rearranging habitats in ways we can't begin to predict, spawning migrants everywhere. We may notice more pine or spruce beetles this year, or fewer familiar butterflies poised like pocket squares atop the flowers, or thinner fire-crisped forests with dusty winds and jaw-dropping heat. We may wonder where all the slender-necked corncrakes have gone. We may obey the rules and not water lawns or wash cars or let the faucet run while brushing teeth. But we may not connect the dots and link less water and missing butterflies and corncrakes to early spring and snowpacks melting too soon, leaving little water for

parched forests during the long torrid summer, when already weak-
ened trees face an armada of beetles and incendiary drought.

This is not so much a vicious cycle as a carelessly torn fabric.
You notice a ripped seam, and though you may procrastinate about
fixing it, it annoys your senses, it picks at your awareness, something
isn't as whole. The foxes have moved north, there are new snakes in
the yard, field mice have either waned or multiplied to Pied Piper
of Hamelin status, West Nile virus is slaying the local crows, and
you spotted something long with eyes and scales swimming in the
canal. Lured by the warmer, north-spreading swamps, alligators
have begun slithering up from Florida into North Carolina. In time,
they may well become native to Virginia, maybe venturing as far as
Virginia Beach, with some trailblazers swimming up the Potomac
to D.C.

One keystone species, plankton, at the heart of the ocean food
chain, tells a tale of the changing times. Tiny shrimplike flagellates
in the trillions, without a thought among them, they're barely visible to us and seem far too weak to act as a keystone, without which
hazel-waved ocean life would collapse. But they are one of the largest biomasses on Earth, drifting everywhere on the currents like
pointillist clouds.

In Arctic waters, where polar bears travel the corridors of sea ice
with their young, resting and hunting, and seabirds nest on icy cliffs,
flying to fish through cracks in the ice, seals give birth and raise their
young atop the floes. Walruses ride on magic carpets of ice to fish
farther afield. With warmer water, there are fewer icebergs where
algae cling, and fewer algae-eating plankton as a result. According
to a recent study published in *Nature*, worldwide levels of plankton
are down 40 percent since the 1950s, which means less food for the
plankton-feeding fish, birds, and whales.

Less plankton leads to fewer krill, tiny crustaceans whose numbers have also plummeted, and fewer petite Adélie penguins, which
feed on krill and squid in Antarctic waters at the other end of the
globe. Untidy masons of the penguin world, Adélies build nests of

stones along gently sloping beaches and raise fluffy, brown, yeti-shaped chicks in those miniature craters. When I visited one large, squawksome colony twenty years ago, stones were a precious commodity. But, according to the ornithologist Bill Fraser, that Adélie population has dropped by 90 percent in the past twenty-five years. With so few couples courting, there are stones abounding, but less food for the orcas (killer whales) and leopard seals that prey on the penguins.

Yet the Anthropocene (like nature itself) rarely tells simple stories. In Alaska, our bestirring of the weather is good for the nearly extinct trumpeter swans, who are using the longer summers to feed and raise their young. Orcas will also profit from the warmer waters. As Arctic seam ice shrinks to a record low, undulating orca shipping lanes open up across the pole via the once-fabled Northwest Passage, changing the ecology of the northern ocean. The melt allows the orcas to widen their range and catch more of the white "singing" beluga whales, the canaries of the ocean, and the unicorn-tusked narwhals, two of the orca's favorite meals. But both the belugas and the narwhals are endangered.

How astonishing it is that just one warm-blooded species is causing all this commotion. Creating hives of great megacities and concrete nests that tower into the sky is impressive enough. But removing, relocating, redesigning, and generally vexing and bothering an entire planet full of plants and animals is another magnitude of mischief beyond anything the planet has ever known. The first is just brilliant niche building, something other animals do on a much more modest scale. For instance, beavers fell trees and dam up streams to create ideal ponds for their underwater huts, and in the process some flora and fauna are dislodged. But no other animal widens its niche to disturb every life form on every continent and in every ocean.

The addled climate is boosting some species and harming or extinguishing others, and not in faraway places, but close to home, in signposts as plain as the jamboree of Canada geese on your lawn.

This news of climate change isn't accusatory, jargon-ridden, argu-able, or even verbal. It's local and personal when eagerly awaited butterflies—the ones that captivated your parents, you, and your children every Good-Humor-jingling summer you can remember—have fled. Some things are more visible in their absence.

In England, the once-rare Argus butterfly has been extending its range northward over the past thirty years, and altering its diet in habitats free of its natural enemies (parasitoids). It's a marvel with brown wing tops fringed in white, and bright-orange eye dots; the underside is paler brown with black and white eye dots plus the orange. The North Country nurse, leaving her local pub, won't see her favorite winged pub-crawlers flitting across the meadows. Until she visits her sister thirty miles farther north, where the rare beauty is now plentiful. *How come you now get the butterflies and I don't?* she may be thinking with a touch of sibling eco-rivalry. Parasitoids used to finding Argus caterpillars on certain plants haven't kept up with Argus's northward migration.

Imagine if you woke up one morning and discovered that the res-taurants across the street, the nearby deli and groceries—in fact, all your usual food pantries—had moved several hours north during the night. Would you make long tiring shopping trips, change your diet, or follow the food and resettle in the north? Like our hunter-gatherer ancestors, you'd probably pack up and follow the herds. (Our herds may lie motionless on shelves these days, but we're still out there hunting and gathering.)

The Spanish ornithologist Miguel Ferrer estimates that around twenty billion birds of many species have altered their migration pattern because of climate change. "Long-distance migrators are traveling shorter distances; shorter-distance migrators are becoming sedentary," he reported at a conference of two hundred migration specialists. "The normal summer temperature in your city twelve months ago is now normal four kilometers further north. It doesn't sound like a lot, but that's twenty times quicker than temperatures changed in the last ice age."

An Audubon Society study found that roughly half of 305 species of North American birds are wintering thirty-five miles farther north than they did forty years ago. The purple finch is wintering four hundred miles farther north. Birds are fun to watch and beautiful, of course, but they're also essential pollinators, seed dispersers, and insect eaters whom we need to help crops and ecosystems flourish.

It's not always easy for birds to migrate in our citified, fragmented landscapes. Some species, fooled by warm weather into traveling too early, arrive in new environs before the food is sprouting in the fields. Once upon a time corncrakes filled the skies from northern Europe to South Africa with long-necked speckled charm and hoarse calls, nesting in field edges, or in dense vegetation and grasslands. As we've mown those fields to plant crops, corncrakes have lost their foothold and become scarce. Fortunately, through the synchronized efforts of fifty countries, from Belarus to Tasmania— such tactics as asking farmers not to mow their grasslands until after the chicks have fledged—corncrakes and their tottery black young are making a comeback.

There's often little harm done by nudging plants and animals to new locales. That's how we (the most successful invasives of all time) came to settle the continents, planting apples, peaches, horses, and roses. Only crab apples are native to North America. The sweet crisp honey-scented apples we wait for all summer, the fleshy Red Delicious we shine on our jeans, and the 7,500 other cultivars, each with its own taste, fragrance, crispness, and uses—all are invasives, and we've bred them like show dogs to taste, feel, look, cook, and smell the way we prefer. Since I'm an apple maven, I'm grateful for two old tart apple trees growing in my backyard.

I also find horses a wonderful addition to the North American landscape, even though I know they're an invasive species that arrived on Spanish ships and adapted to the vast grassy habitats. It's a mystery how this happened, but if we look at one instance—the wild ponies on the East Coast barrier islands of Chincoteague and Assateague—there are intriguing theories. Did pirates turn horses loose to graze while they were off looting, and return to find they'd vanished into the dense thickets and woods? Did seventeenth-century growers, who imported European horses, decide to pasture them on the island to avoid taxes, and discover that some wandered off to begin a herd? Did a Spanish galleon, bound for the English colonies, encounter a hurricane and break up on the island, where local Indians came to their rescue, but the ponies ran free? Did a boatload of Spanish horses, blinded for work in the mines, sink in a hurricane, and the terrified horses, despite the raging storm and their blindness, somehow manage to swim to shore? No one knows.

The ponies' ancestors faced poor food, scorching summers and cold damp winters, sandpaper winds, relentless mosquitoes, and cyclical storms. In response they grew thick furry coats and learned a host of survival skills, such as sensing a drop in barometric pressure, seeking shelter in hilly areas, and huddling together with their rumps facing the high winds. As a result, only the fittest and smartest ponies survived, and their genes live on in the current herd, which is vigorous, canny, and well adapted to the rigors of a maritime landscape.

I may be able to trace my beloved rosebushes, which I raise organically, tend minimally, and almost never count, back thirty-five million years to fossils, or five thousand years to the Chinese gardeners who first dreamt of breeding them. They're not native to North America either, but also highly successful invasives, which ran wild to the surprise and merriment of settlers.

Chinese mitten crabs may be destroying the San Francisco Bay habitat, but European green crabs are helping to revive the ecosystem in the salt marshes of Cape Cod. We're restoring prairie ecosystems by reintroducing the grasses that used to fan slowly in the summer haze, and reclaiming wetlands by diverting streams and planting native flora. When we do, of course, it changes the climate and the migration patterns of birds and insects; some animals find a new home, while others decamp. In most cases, the ecological restoration is successful, but we sometimes get it wrong.

The southernmost of the Mariana Islands, about 1,500 miles south of Tokyo, Guam has been a U.S. dependency since the Spanish–American War of 1898. People think of it mainly as a military base, but it is also a lush tropical paradise. Yet at dawn all a visitor hears is an eerie silence, no birdsong, because most of the native birdlife has gone extinct. What's been causing the extinction? Pesticide poisoning? Habitat destruction? Exotic disease? It took scientists years to figure it out. The answer is one pregnant Indonesian brown snake that slithered off a cargo ship in 1949. The nocturnal brown tree snake is an arboreal predator, up to eleven feet long, and poisonous. Equally happy in forests and on rocky shores, today the snake has multiplied into the hundreds of thousands. Birds evolved on Guam with no predatory snakes, and so most of the forest birds have been devoured, along with native reptiles, amphibians, and bats. The flightless Guam rail survives only in captivity, rescued in a last-ditch effort to save the species by starting a captive breeding program on Guam and in some mainland zoos, so that it doesn't go the way of the Guam flycatcher, a species once unique to the island. Because birds were the chief seed-spreaders of native fruit trees, those are vanishing, too, and the island is aswarm with forty times more spiders, which the missing birds used to hunt.

The same fiddling with evolution can happen on much larger islands. In Australia, cane toads—native to the Americas, from south Texas to the Central Amazon basin—were imported as assassins to eat the plague of cane beetles, which they did with rousing success. In their native habitat, cane toads can grow as big as catchers' mitts and weigh five pounds. In their new home, they also evolved longer legs to cross the vast outback. Only big-mouthed snakes could swallow the poisonous toads, and those snakes died in the process, leaving their smaller-mouthed, toad-shy cousins to pass on their genes. As a result, Australian snakes began evolving smaller mouths.

Sometimes we breed and move animals around the landscape to save their species from extinction. In *Wild Ones*, Jon Mooallem writes about his experience with Operation Migration, where endangered baby whooping cranes are hatched in incubators and taught wild crane behaviors, including how to migrate, and then are led thousands of miles by crane-suited humans flying ultralights. "This work," Mooallem writes, "this wholesale manufacture of wild birds by human beings—turns out to be so ambitious, tedious, and packed with perplexing arcana that, after ten years, it's hard for those who have given their lives to the project even to agree on how well it's working and what they should do next."

Usually, however, we breed and transport animals not to save them but to suit ourselves: domesticating dogs, cats, and even pythons as pets, enslaving horses and oxen for work, or drafting a cavalcade of creatures—from camels and horses to pigeons and bats—to fight beside us in our bloodiest wars.

"THEY HAD NO CHOICE"

When the life-size horse puppet first appears onstage in *War Horse*, its neck rippling and ears twitching in startlingly lifelike ways, it takes a moment to figure out what humans are doing inside a see-through horse. But as the puppeteers brilliantly animate the tendons and muscles, it makes sense to the inner shaman inside us, the being who identifies so closely with animals that we've embedded animal attributes into our slang (eagle-eyed, stubborn as a mule, lionhearted, strong as a bull, etc.). All children play at being other animals. We often wallpaper nurseries in animal motifs. We use animals as avatars online, and we become part animal in horror movies, not to mention soap operas that feature love-sick half-animal vampires. In the past, we attached our muscles to those of animals, who extended our speed and strength in battle, or carried our supplies. They became a kind of equipment that men often grew fond of, and yet had to watch die in battle, or leave behind at war's end. As we've been learning more about the minds and senses of animals, discovering that they experience emotions similar to our own, we're developing much more compassion for them.

One moment in the middle of Steven Spielberg's lyrical epic film version of this story still haunts me. As we are charging through the woods during World War I, amid gunfire, shrieks, and bloodletting, with young men and horses shell-shocked and crazed by horror, the action recedes for a needed break from battle, the camera pauses, and we view the war in a surprising and intimate mirror—reflected on the curved eyeball of Joey, the equine hero. Trauma, like some hallucinogens, lingers for a long time in the tissues, and in that shot one can see exactly how it gets there. Because a horse's eye is curved, the scene is warped, with men and animals and smoke and balls of light and flying clots of earth leaping in all directions. War is traumatic for horses and other animals—like the humans who create it. The beautifully wrought image says it all, and stings the heart like a line of saber-sharp poetry.

We've recruited many unlucky other animals to fight our battles, and only recently have we begun to recognize their capacity for horror and suffering, and to commemorate their sacrifices. Between two busy streets, near Hyde Park, in London, I happened upon a startling war memorial surrounded by a lively crowd of people, several horses and mules, a handful of well-behaved dogs and cats, and a flock of racing pigeons. Older gents in military caps, some in bright regalia from wars long past, displayed medals on their chests and red poppies in their lapels. A mounted Household Cavalry soldier, dressed all in black, rode a matching jet-black Irish horse, which he seemed at times to fade into, except for his white belt and the scarlet stripe down his pant leg and the scarlet band around his cap. Other soldiers in desert camouflage, well-dressed women holding lapdogs, countless veterans, animal rights groups—all milled in admiration around the gleaming sixty-foot curve of white limestone symbolizing the arena of war.

Beautifully rendered in bas-relief on the wall, a parade of camels, elephants, monkeys, bears, horses, pigeons, goats, oxen, and other animals bravely march side by side to war. A few yards away, two heavily laden bronze mules, their strength whittled to the bone,

struggle up shallow steps under a burden of rifles and battle equipment, heading toward a breach in the wall. The first mule stretches
its arrowlike neck toward a garden visible through the pearly white
gates of limestone.

On the other side of the wall, a robust bronze stallion is breaking
into a gallop, with a bronze setter beside it, both freed of their bur-
dens. The dog's head is turned to look back toward fallen comrades.
Life's bas-relief gone, on this side of the wall animals are carved in
stark silhouettes and hollowed outlines, like a child's puzzle waiting
for the pieces to be slipped in.

The £1 million memorial, paid for entirely with private funds,
bears this legend:

> *This monument is dedicated to all the animals that served and died
> alongside British and Allied forces in wars and campaigns throughout
> time.*

A smaller inscription beneath it reads:

> *They had no choice.*

I heard several old soldiers paying tribute to the animals they
relied on in so many ways during the war. A veteran sang the praises
of the supply mules in the Burmese jungle (their vocal cords had
been cut lest they bray and endanger the soldiers). "My life was saved
by the mules," he said, his gaze sliding into the past. "The only way
we could get the guns up to us was using them." Some anonymous
animal lover had left a wreath on the memorial whose card read:
"You have smelt our fear. You have seen our bloodshed. You have
heard our cries. Forgive us dear animals that we have asked you to
serve in this way in war."

Nothing prepares you for how cold the bronze mules are to the
touch, or the camel's sand-matted cement coat. In springtime, beds
of daffodils are blooming yellow trumpets. Behind the curve, in the

afterlife part of the memorial, the horse is much larger than life, its hooves the size of dinner plates. Too tall to mount, it walks on a lawn of grass sprinkled with tiny daisies. The setter is life-size. Neither has been endowed with a human expression. They are simply and amply animal, healthy and untroubled, relieved of war's horrors.

Designed by David Backhouse, the memorial poetically captures the plight of the millions of animals who have served and died in our wars. Dogs carried reels on their backs and laid telegraph lines, or ripped their paws raw digging through rubble for survivors. Orca whales became cinematographers, patrolling while holding cameras in their mouths. Pigeons delivered messages from the front. Sea lions dived to 650 feet to recover lost equipment. Beluga whales learned to dial their sonar for surveillance in waters too cold for other mammals to dare. War elephants were ridden on campaign through mountains and jungles. Camel cavalries battled in Arabia and North Africa. And glowworms . . .

Yes, glowworms. In separate chambers of their body, glowworms (also known as lightning bugs) brew luciferin and luciferase, two chemicals that don't do much until they're mixed together. Then they become a magic potion that glows so brightly the insects can use it to blink semaphores of love. Their hind ends become literal lighthouses, leading mates to shore. Sometimes "femme fatale" lightning bugs interfere, by mimicking another female's flash code and stealing her mate. A siren's come-hither can be mesmerizing, even on the battlefield, and fill the heart of an insect suitor with light or illumine a soldier's letter from home. Unlike an incandescent bulb, the light blends into the landscape. And so, during the trench warfare of World War I, soldiers of the Somme read their maps and letters by the cold green light of glowworms carried to war as living lamps.

A hundred thousand pigeons flew missions during World War I, and two hundred thousand during World War II, racing a mile a minute to deliver strange cargo—coded notes in capsules taped to their legs. One notable World War I pigeon, Cher Ami, who flew for the U.S. Army Signal Corps (and died of battle wounds in 1919), relayed twelve urgent messages before being shot in the breast and leg. Despite the blood loss, shock, and shattered leg, he delivered his message—an act dubbed "heroic" by the French, who formally awarded him the Croix de Guerre. Trained at Fort Monmouth in New Jersey, and recruited as an avian soldier, Cher Ami could hardly be thought of as "serving his country." Still, Cher Ami's one-legged body is on display in the Smithsonian National Museum of American History's *Price of Freedom: Americans at War* exhibit.

Stranger by far, at the height of World War II, the Americans prepared Pacific-bound Project X-Ray, also known as the Bat Bomb Project, dreamt up by Lytle S. "Doc" Adams. Bats had always played an important role in U.S. warfare, because bat guano ferments into saltpeter, a key ingredient in gunpowder, and as early as the Revolutionary War soldiers scraped and mined it from caves favored by migrating bats.

Gathering thousands of Mexican free-tailed bats from Bracken Cave, where twenty million mom and baby bats roost between March and October, Adams and his team fitted them with small incendiary bombs, and planned to tuck them into individual canisters, each equipped with a parachute, and drop them over Japan. At a thousand feet the canisters would open and the bats fly free to roost under shingles and eaves, where they would soon explode, setting fire to whole cities dotted with wooden and paper houses. President Roosevelt okayed this oddball plan and spent $2 million on it. Then one day armed test bats accidentally escaped and torched a Texas air base, after which Project X-Ray was ditched.

Also at work during those years, the American behaviorist B. F. Skinner began developing a pigeon-guided missile. In what was known as Project Pigeon, he trained the birds to steer by pecking at a target. The U.S. Navy revived it after the war as Project Orcon (for "organic control"), and only abandoned the scheme in 1953 because electronically guided missiles proved more reliable. It wasn't declassified for another six years, just in case.

It's well known that pigeons, dogs, horses, camels, and elephants have been drafted for war since ancient days. Apparently pigs served in battle, too. Pliny the Elder, who lived in ancient Rome, tells of herds of grunting hogs being loosed to scare the elephants of invaders. Lately, though, we've extended the idea of animal soldiers into the realm of lunacy.

The Pentagon's Defense Advanced Research Projects Agency (DARPA)—whose "psyops" (psychological operations) division most people first learned of in the book and film *The Men Who Stare at Goats*—trained CIA operatives to practice remote killing on animals. But that was only one of the CIA's bizarre plans for animal combatants. In Operation Acoustic Kitty, the CIA implanted a bugging device inside a cat, with an antenna hidden in the cat's tail. This five-year-long, $5 million Cold War project was canceled when the cat, released near a Russian compound, was hit by a car and died while trying to cross a street. Cats were also considered as guidance systems for bombs dropped on ships (the tenuous logic being that since cats hate water they'd steer the bombs they were strapped to toward the deck).

Animals have died in the millions helping us fight our wars, and as our soldiers have become increasingly more technological in recent decades, so have our armed service animals. In 2010, the Chinese newspaper *People's Daily* accused the Taliban of training monkeys to shoot Kalashnikovs, light machine guns, and fire mortars at NATO forces. Though the Taliban denied the rumor, it leaves disturbing images in the mind, and conjures up the scene in *The Wizard of Oz* when a battalion of flying monkeys attacks from the skies. Even watching the movie as a kid, I was scared not by the monkeys but by the witch evil enough to train animals as goons to wage our wars.

The CIA has experimented with remote-controlled cyber-insects, inserting microchips into the pupa stage of butterflies, moths, and dragonflies because, as a DARPA proposal explained, "through each metamorphic stage, the insect body goes through a renewal process that can heal wounds and reposition internal organs around foreign objects." The result: cyborg dragonflies and robomoths, and search-

and-rescue cyborg cockroaches. Other plans have included remote-controlled sharks (with electrodes in their brains) designed to sniff out bombs and explosives, bees trained to replace bomb-sniffing dogs, and hamsters stationed at security checkpoints who are trained to press a lever when they smell high levels of adrenaline.

For more than fifty years, the U.S. Navy has trained pods of dolphins to use their elite echolocation skills and low-light vision to spot and clear underwater mines. They've served the navy in both Vietnam and the Persian Gulf, by filming, delivering equipment, and capturing enemy divers (by clamping on a leg cuff roped to a buoy). Mine-hunting dolphins learned to identify underwater explosives without detonating them and report back to their handlers, giving yes or no responses to questions. Sometimes they marked the whereabouts of mines by delicately attaching buoy lines to them; other times they disabled the mines by attaching explosives and dashing away. When Iran threatened to mine the Strait of Hormuz in 2012, and block the vital shipping route, NPR asked retired admiral Tim Keating, who commanded the U.S. Fifth Fleet in Bahrain, how he'd handle the situation.

"We've got dolphins," he said matter-of-factly.

Prized as the dolphins may be by their handlers, the navy regards them as another form of personnel, without rank but classified in true military fashion. For instance, "Mk 4 Mod 0" is a dolphin trained to detect a mine near the seabed and then attach an explosive charge to it; "Mk 5 Mod 1" is a sea lion used to retrieve mines during practice maneuvers.

However, in the summer of 2012 the navy made a momentous announcement. It plans to retire its squads of minesweeping dolphins and other sea mammals by 2017, and replace some with robotic drones called "knife fish" (a fish known for emitting an electric field). Unlike sea mammals, knife fish won't be able to neutralize a mine, only locate, film, and transmit data about it. Other pods of robotic underwater drones will be guided by fiber-optic cables. I'd like to believe that compassionate motives inspired this decision, but

I'm sure thrift also played a role. As a military system, robots are cheaper to field than dolphins, who are heavy to transport to battle theaters in water-filled tanks, and require feeding and medical care.

There's been a lot of public complaint about sending animals to war and research labs, especially big-brained mammals like dolphins, and maybe it hasn't fallen on entirely muffled ears. In 2013, after years of concerted worldwide lobbying, the U.S. government finally retired nearly all research chimpanzees and listed the species as endangered. This means that the chimpanzees who have weathered countless illnesses on our behalf are finally being released to animal sanctuaries, and no new chimps will have to face such horrors.

Our dominion over animals, ill-treated for eons in our research and wars, is rapidly being replaced by technology, thank heavens. We have pack mule robots designed to supplant horses and trucks in difficult terrain, robot fleas that leap through open windows and spy, and ambidextrous gymnastic robots that can fill in for human soldiers in toxic areas. But no one has figured out yet how to engineer a dog's superrefined nose. Dogs can smell a man's scent in a room he has left hours before, and then track the few molecules that seep through the soles of his shoes and land on the ground when he walks, over uneven terrain, even on a stormy night. Thus far a robot can't match that finesse. So for the time being we still have dog-soldiers, some trained to kill.

If wars must be fought at all, drones at least are heartless. Alas, their targets aren't.

PADDLING IN THE GENE POOL

For thousands of years, we've left a trail of our preferred traits in the planet's life forms, in food crops and animals, to be sure, but also in our favorite pets. No other animal is as defined by the long history of its codependency with humans as the dog, so much like us in its capacity for affection and savagery. All dogs trace their ancestry to one canine—the wolf—but you'd never know it looking at Chihuahuas, Bedlington terriers, Belgian griffons, cocker spaniels, Great Danes, boxers, Basenjis, Afghan hounds, or corgis. Over the centuries, we've bred dogs for all sorts of jobs and sports: long dachshunds to squirm down badger holes, balloon-chested greyhounds and whippets to race, Entlebucher Mountain Dogs to herd sheep. Lapdogs abound in a fantasia of shapes and colors. Or one can choose a companion dog by intelligence, disposition, or possibly endearing neuroses (tail-chasing, for example). Our feats of selective breeding produce trendy, aesthetically pleasing dogs, including many so inbred that they suffer from about 350 known hereditary diseases, including beagles with weak spinal discs, Dobermans given to narcolepsy, basset hounds with blood-clotting woes, flat-faced Pekingese bedeviled by breathing problems, and Scottish terriers eighteen times more likely to develop bladder cancer. We design dogs so small that they tend to dislocate their kneecaps, and dogs so large they have trouble with their hips.

But, however misbegotten the result of retooling our fellow creatures by controlled breeding, it is an experiment we've been carrying on for so long that we no longer even recognize the results as "unnatural." Now that we have the technology to reach a steely hand into the machinery of cells and remodel the genes, even human ones, our powers are more disturbing, raising all kinds of ethical and legal challenges.

In 2012, John Gurdon and Shinya Yamanaka shared the Nobel Prize for the breakthrough discovery of how to persuade adult skin cells to regress into jack-of-all-trades ("pluripotent") stem cells capable of morphing into any type of cell in the body—heart, brain, liver, pancreas, egg. It's as if Gurdon and Yamanaka had found a way to reset the body's clock to early development, enabling it to mint wild-card cells that haven't chosen their career yet—without using the fetal stem cells that cause so much controversy.

Space may be only one of the final frontiers. The other is surely the universe of human imagination and creative prowess in genetics. "We are as gods and might as well get good at it," Stewart Brand began his 1968 classic, *The Whole Earth Catalog*, which helped to inspire the back-to-the-land movement. His 2009 book, *Whole Earth Discipline*, begins more worriedly: "We are as gods and *have* to get good at it."

Among the rarest of the rare, only several northern white rhinoceroses still exist in all the world. But, thanks to Gurdon and Yamanaka, geneticists can take DNA from the skin of a recently dead animal—say, a northern white rhino from forty years ago—turn it into "induced pluripotent stem cells" (IPS), add a dose of certain human genes, and conjure up white rhino sperm. Then, using in vitro, they hope to fertilize eggs from a living female white rhino, producing offspring to add genetic diversity and rescue the species. Thus far, they've successfully created embryos.

With vase-shaped dark faces haloed in soft silver fur, as if even their foreheads were bearded or their auras were permanently visible, silver-maned drill monkeys are the most endangered African primates, swinging through tiny swatches of jungle in West Africa. The few survivors are riddled with diabetes. As a joint project, the San Diego Zoo and the scientists at the Scripps Research Institute are hoping to find a cure for the disease and also enrich the drills' gene pool: they recently created IPS from the silver-maned drills and coaxed the stem cells to become brain cells.

The same brew of magicians at Scripps is hatching IPS therapies for humans, with the first long-awaited human clinical trials for Parkinson's disease starting soon. That's a far cry from dinosaurs grazing in Central Park. Not that dinosaurs are on the table; they've been extinct so long that their DNA is no longer viable. But lots of other animals, from passenger pigeons to great auks, are contenders. Curiosity is a natural torrent in humans, one it's hard as a mudslide to resist. Hosts of people meet avidly at conferences and in research centers to discuss who, how, when, and why to de-extinct. Russia's Pleistocene Park, where look-alikes of ancient cattle and horses already roam, awaits a shaggy herd of woolly mammoths, even if the grassy steppes they once grazed no longer exist. For some, de-extinction is a moral quest dusted in eco-guilt; for others, it's a burning scientific challenge. A candid few openly admit that, for them, it's just too cool an idea to pass up.

At Harvard, the molecular geneticist George Church has pioneered ways to ramble through orchards of DNA, cherry-picking individual genes to produce desired traits or remove harrowing ones. I've always found Tasmanian devils endearing, maybe because they're raucously bad-tempered and quarrelsome to a laughable extreme, and stand up like hairy sumo wrestlers when they fight. Their wild population has been ravaged by contagious facial tumors, and about 80 percent of them have died. Church could pinpoint the gene that's causing the tumors and erase it from the bloodline. He could recreate extinct passenger pigeons by cobbling together bits of their DNA, if he wished, writing a formula using DNA's four bases: A, T, G; and C. Despite the futuristic science and lab settings, the vocabulary he and his cohorts use is mechanical and pure Industrial Age. To build a cell, a bioengineer consults a registry of parts, chooses the bio-bricks he wants, and goes through an assembly process: loading bricks of DNA onto a chassis (the *E. coli* bacterium, for example) in a foundry.

It's astonishing that we've come far enough as a species to think, *We've driven all these animals extinct . . . how can we restore them?* We have a sense of deep nostalgia about the animals that surround us, and the possibility of de-extincting them. We used to think we had simple dominion over the animals. Today, as we watch plant and animal gene pools dwindling, and pluripotent stem-cell technology zooming, we know that our role is far more complex and that ingenious mistakes require even more ingenious responses.

AS THE EAST Midlands train glides from London toward Nottingham, we pass a show-stopping array of giant old horse chestnut trees with domed crowns and wide shady skirts. Covered in cone-shaped blossoms that sway like incense, they scent the air with a heavenly smell that's fierce as lilac but more animal. Not barnyard horsey exactly, but leather-sweet and slightly sweaty. Soon the suburbs give way to brilliant yellow stripes: fields of flowering rapeseed whose oil once lubricated our heavy machinery and now enters our trucks as biodiesel and our gullets as canola oil. Around a bend, tall termite mounds sprout in the distance, but as we draw closer they become smoke-gushing volcanoes, and then finally loom as the seven cement chimneys of a coal power station. Clustered together, they're a monument to the Industrial Age, and the tallest landmark (some say eyesore) for many miles. For two hours, we've been escorted by an endless tribe of metal stick men, each with three arms from which steel pinecones dangle, who stride across the land holding up wide skeins of power lines.

All this sparsely settled farmland is what we call the "country-side," and it fills the eye with bucolic English grandeur, even though some of it is far from native (Canada geese), some has been geneti-cally engineered (canola rapeseed), and much has been machined from lyrical spurts of steel. Even the horse chestnut trees, so syn-onymous with England, have Balkan ancestry and an industrial secret. During World Wars I and II, British children collected their seeds—the deliciously smooth conkers that we carry in our pockets and serenely rub, polishing them with body oil until they look like mahogany knobs—and donated them to the war effort, where they helped to brew cordite for explosives.

No clouds of starlings pepper the sky before turning on a knife-edge and circling round. The overuse of pesticides has silenced the wildlife they feed on, and now great flocks and swarms and slith-ers of animals have vanished. Ironically, as a result of crops blanket-ing 70 percent of the United Kingdom's land, more biodiversity has begun haunting the cities. Friends tell me London is askitter with red foxes.

I'm struck by the uniform, yielding, furrowed richness of the miles. Single crops parade past the window, along with clans of cattle and sheep. The same monotony of genes rules much of the planet, as wild and varied habitats give way to more prosperous if homogenous big farms. In the United States alone, twenty thousand square miles of land are covered by corn—an area twice the size of Massachusetts. We're dabbling in eugenics all the time, breeding ideal crops to replace less aesthetic or nutritious or more perishable varieties, leveling forests to graze cattle or erect shopping malls and condos, planting groves of a few familiar trees homeowners and industries prefer. In the process, there's the gradual eradication of genes, without fanfare, sometimes even driven by good motives.

We're at a dangerous age in our evolution as a species: clever, headstrong, impulsive, and far better at tampering with nature than understanding it. Who knows what vanishing life forms—and their DNA—we may one day regret losing? Pollen from this sameness of crops will show up in the fossil record as a curiosity of our age; examining it, Olivine will wonder why we wiped out a beneficial smorgasbord of plants in favor of a few genetically modified, fruitless varieties. Will she guess that most of these were bred in the fields of avarice (forcing farmers to keep buying a company's seeds)?

We're in the midst of the planet's sixth great extinction, losing between seventeen thousand and one hundred thousand species a year. It's hypothesized that what caused the first great extinction, 440 million years ago, was radiation from the collapse of a massive supernova. During the second extinction, 245 million years ago, a possible meteor strike combined with volcanic eruptions killed off so many ocean species that coral reefs vanished for 10 million years. About 210 million years ago, some catastrophic event wiped out more than half of all life forms. The extinction that ravaged the planet 65 million years ago polished off the dinosaurs, sent temperatures soaring by nearly 60°F, and pushed sea levels up more than nine hundred feet. Today's extinction event, the first during our reign, could end up being the most catastrophic of all. Many scientists predict that, at the pace we're going, about half of all the world's plants and animals will vanish by 2100. But, for a change, we know the exact causes of the extinction, having created them ourselves—climate change, habitat loss, pollution, invasive species, big agriculture, acidifying the oceans, urbanization, a growing population demanding more natural resources—and we're in a position to stop them, if we set our collective mind to it.

So, as species dematerialize around us, worldwide efforts are under way to collect and protect the DNA of as many as possible before it's too late. Two brave doomsday efforts have been leading the way. One is the Svalbard Global Seed Vault, a remote and heavily guarded underground cavern tucked four hundred feet inside a sandstone mountain on the Norwegian island of Spitsbergen, in the secluded Svalbard archipelago, which lies about eight hundred miles from the North Pole—a James Bond destination safe from both man-made and natural disasters, even melting ice caps (it's 430 feet above sea level), tectonic activity, or nuclear war.

In some climate scenarios, due to crop shrivel and shortage of fresh water, food becomes scarce. The vault's mission is to preserve seed diversity, because one never knows what calamity might descend, which crops might fail, and what seeds—whether heirloom or genetically engineered—might offer a solution in the Anthropocene world. The key may be a relic from a bygone era. As Paula Bramel, assistant executive director of the seed vault, explains: "The environment is changing to the point where farmers can no longer maintain the seeds of the varieties that they always used. And that's really a loss to everybody because that variety may have a trait that's really critical in the future."

If those seeds aren't cached now, they won't sprout later. Bramel points out this is a particular problem in Africa. We've abandoned many heirloom crops because they're not as cheap to grow in our technological Valhalla, or as blemish-free as we demand. Some are no longer as hardy in our changing climes. We may one day wish we still had them. The vault is the Norwegian government's farsighted gift to the world; there's no charge to countries for storing, cataloging, and overseeing millions of backup seeds, which, for the most part, sleep at -0.4°F, in refrigeration naturally provided by the permafrost. Every decade or so, some of the Sleeping Beauty seeds must be awakened and allowed to bloom so that fresh seeds can be collected. Of course, in the best possible future, if such a state of grace exists, they won't be needed. In the meantime, as a fail-safe treasury, the vault houses millions of specimens of over four thousand different species.

The second doomsday effort lives on the campus of Nottingham University, in the Frozen Ark, which stores the DNA of 48,000 individuals from 5,438 different animal species. It's Noah's Ark moored in Robin Hood's backyard, and its logo is a blue sketch of an ark sailing on a double helix of ocean waves.

FOR LOVE OF A SNAIL

Bryan loved snails, though not all snails equally. He was especially fascinated by one turban-shaped gastropod in the genus *Partula*. Among partulas, he had a thing for the species *Partula mooreana*. Just like Darwin's finches, drifting colonies of partulas on tiny islands became cut off from their neighbors and rapidly diversified to a surprising degree in color, size, and shell motifs. Free of natural predators—except the Polynesians who strung ceremonial necklaces from their shells—the snails hoovered up algae on the undersides of caladium, plantain, dracaenas, turmeric, and other leaves in the understory of heavily forested volcanic slopes.

In Bryan's pale-blue office, I'd seen an old book on his desk, opened to favorite pages of colorful plates. Some of the shells were straw yellow and girdled by two or three narrow reddish-chestnut bands, others light-brown-capped with darker whorls, or cream and purple. Despite their different patterns and hues, they were all partulas. Beside it, several snail-shell necklaces lay in casual loops of pink, brown, tan, and gray. Each tiny shell had been carefully matched for colors and size, and what makes this remarkable is that, according to Bryan, even in one valley shell patterns can vary about every twenty yards. It was disturbing to see the remnants of so many tiny lives ghosting across his desk, their sticky inhabitants long extinct. Their spiraling interiors belong in a church designed by Gaudí; their eye-catching shells remain to delight and haunt us. But their empty beauty tugs at the mind.

Once, a hundred or so species of partulas inhabited French Polynesia, where the shells of the most attractive species, brown-and-green-striped, were prized. They bore such incantatory names as *Partula dolorosa*, *Partula mirabilis*, *Partula solitaria*, and *Partula diaphana*. In the 1880s an enterprising French customs officer, who enjoyed the taste of snails, decided to start a snail farm and flog snails to islanders. The snails he chose to breed were the plump, succulent giant African snail *Achatina fulica*. When his snails didn't catch on as cuisine, he tossed them into the wild, where they began raiding the local crops and gardens and hitching rides to other islands. Faced with heavy crop damage, U.S. authorities on Guam introduced a particularly voracious predator, the carnivorous Florida rosy wolfsnail, *Euglandina rosea*, into one orange plantation in 1977. But even snails can be picky eaters, and unfortunately the rosy wolfsnail didn't find giant African snails tasty. Instead it dashed into the adjoining forests, where it hunted down and feasted on the partulas, which for some reason it fancied. Online one can watch ghoulish footage of the carnivores at work, plunging into the shells of small, helpless partulas and gulping them down in mouthfuls with cannibalistic gusto. Ten years later, these rosy wolfsnails commanded the whole island and had devoured fifty species of partulas. Yet another tragic tale of invasive species (and human intervention) gone awry.

Volcanic archipelagos offer natural laboratories for studying how species evolve and reinvent themselves, and so, in the 1990s, Bryan Clarke, his wife, Ann, and their lab assistant, Chris Wade, traveled to French Polynesia to study the partulas. As it happens, I was there at almost exactly the same time; though I was not in pursuit of snails, I know the tapestry of sensations they would have worked among:

The spicy sweet smell of scorched sandalwood wafting through the air. Six-man outriggers pulling swiftly past, each with a piece of tusk-shaped wood tied to one side for balance. Dogs sleeping under over-turned outriggers in a budget of shade. The sky everywhere full of seabirds, and on land pairs of fairy terns perching like small white angels among the tree limbs while balancing their single egg on a branch. Men heavily tattooed, women swaying as they sang tradi-tional songs filled with baying and keening. Small houses lining the village roads. Here and there, in the foothills, house lights sparkling from the foliage, and beyond them, raincloud-wrapped mountains steaming like volcanoes. Elaborate designs everywhere one looked—on church stones, wooden carvings, and bark cloth—curving like vines and whorling like snail shells, worlds within worlds. It was as if the native artisans had looked through microscopes into the heart of cells.

As a tantrum of sun flashed across the cobalt-blue sea, they boarded a long red-and-white ferry, lashed to the dock by four thick ropes like a wild animal that might otherwise escape, and traveled among small islands with hillsides of bottle green and slate cliffs plunging straight into the sea. It was in this remote spell of thick foliage, light-years from the clipped lawns of England, that they des-perately searched for the tiny, rare, beautifully ribboned tree snails.

They found a handful of live partulas on the island of Moorea, but Bryan had a hunch that the ancestor of all 126 partula species might live on the lone island of 'Eua, about two and a half hours from Tonga. Thirty or forty million years older than the other islands, 'Eua isn't volcanic but a flat chunk of Gondwanaland shelf that broke off and plunked itself next to Tonga. Would they find there the ur-snails that populated all of the Pacific islands?

When they arrived, they discovered to their dismay that the 'Eua islanders had chopped down all of the rainforest in order to plant farms of manioc. Partula's habitat was gone. Only one faint possibil-ity remained. At the very bottom of 'Eua's steep ravines, a thin line of trees nestled beside streams. Islanders didn't risk the climb, and neither did Bryan, but Ann and Chris had come too far to turn back. Pressing on, they slogged down the gorges into the island's deep green pockets. After three weeks of scouring the plateau, beaches, and the last shreds of rainforest on the island, all they found was one midden of empty shells in a stream at the bottom of a crevasse. Picking up the small, round shells, and turning them gently like ancient coins, they realized that they were about twenty years too late, since snail shells last only twenty years in the wild. Grief-stricken, they held the remains of an extinction in their hands.

According to Ann, the combined loss felt too heavy to bear. Carefully packing a few of the live *Partula mooreana* they'd found into lunch boxes, they journeyed home to Nottingham. Most of these snails they shared with the London Zoo, which set up a captive breeding program, and Bryan and Ann began breeding the rest, aided by a technician who was a genius at caring for partulas and getting them to have babies. Unusually, even for snails, partulas don't lay eggs, but have live babies, complete with shells on.

"Here's a partula with young," Ann says, opening a book to the startling image of a snail with a perfectly formed second snail emerging from behind its head (where the reproductive organs hide). "These get born just like that."

"With a soft shell, surely."

"No, with a *hard* shell. They're born as complete adult snails."

Curiouser and curiouser, I think. The tiny necklace-worthy striped snails that haunted the trees in the kingdom of Tonga sprang from the sides of both hermaphroditic parents fully formed with (surely this birthing might hurt?) hard calcium shells. Not to mention, during courtship they stab each other in the head with calcium "love darts" pulled from their quiver.

Aghast at the pace of Partula's extinction, a handful of zoos worldwide also began breeding the snails with some success, from a fenced-in partula preserve on Tahiti to the Snail Room at the London Zoo, where one can see them, small as a fingernail, slow-motion slime-skiing along the glass with their nether parts exposed.

But there's no use releasing them back into the wild while hungry packs of rosy wolfsnails still rampage.

"When we came back with the partulas in lunch boxes we were very despondent," Ann says wistfully, "and I think it was at that very moment Bryan and I thought of the Frozen Ark. Because we'd seen all these partulas dying out. We thought, well, we're doing this for the snails, who else is doing it for other endangered species? We started to hunt about and we really couldn't find anyone."

The Clarkes set up the Frozen Ark Project in 1996 as a response to this crisis, with a single simple objective—to save samples of frozen cells containing DNA from endangered animals before they go extinct. Not as an alternative to preserving animals in their natural environments or to keeping them in zoos, she stresses, but as crucially important extra insurance.

Today a consortium of twenty-two of the world's finest zoos, aquariums, museums, and research institutions have climbed aboard the Ark and are providing DNA. Only a tiny dab is needed, gathered painlessly from mouth swabs, feces, hair, feathers, or blood during routine veterinary visits. The samples are sent to Nottingham, where they're cataloged and safely frozen, with backups stored at the home zoos. It's convenient that DNA is thousands of times smaller than a gnat's whisker; many individuals exist only as wispy smears on small white filter papers. A single-car garage could store a million; a briefcase could hide enough to repopulate a continent. At the Frozen Ark, specimens hover in liquid nitrogen at -196°C, ensuring viable DNA for hundreds of thousands of years, and hundreds of years for complete cells. Nothing moves at -196°C, but in time these cells could be resurrected and recultured.

Using Gurdon and Yamanaka's pluripotent stem-cell work, the Clarkes discovered how the Ark's frozen cargo could repopulate extinct or nearly extinct herds.

"It means we can make any tissue, including eggs and sperm," Bryan says. "Now, the importance of that is extraordinary, because you could, in principle anyway, reconstruct an entire organism, even when it's gone. The Japanese, among others, are trying to implant a woolly mammoth embryo into an elephant which would then give birth to it. There are a lot of mammoths frozen in the permafrost. . . . Of course not for long, because it's melting."

Even after ten thousand years of icy slumber, these mammoths are still a treasure trove of frozen DNA, and Bryan assures me that they'd survive freezing for even longer periods of time, and still be viable, if the right method was used.

"I'm very keen at the moment on the idea of freeze-drying cells, you know, like your coffee."

I try not to picture a jar of freeze-dried woolly mammoth crystals on a shelf beside a jar of dodo crystals, rather like in a country store, but it's no use.

"The habitat up there in Siberia might be just fine." Bryan settles back deeper into his chair. "The question then is: What do you do with the mammoth when you've got it?! That's a sort of quandary. Is it better to have two mammoths so they can reproduce and restore the species? I think that would be fun," he says, clearly charmed by the idea, "and I think they would find habitats they can live in."

It's delightful to imagine twenty-first-century woollies, born from elephant mothers (rather surprised at their shaggy offspring?), thunder-stomping through Siberia, maybe sounding an alarm when startled by a de-extincted saber-toothed tiger.

Yet, as we know, mothers teach their babies all kinds of things; our newborn woollies wouldn't have a mammoth culture. Perhaps they'd imprint on their new mothers like baby ducks and adopt the elephant's ear-flapping semaphore and yen for dust-baths. Or would they be raised like endangered baby sandhill cranes, who are fed and taught crane behaviors by white-costumed humans flying ultra-lights? So they'd be woolly mammoths, but not exactly. These mammoths would probably also not have the ancient ancestral suite of woolly mammoth bacterial DNA in their gut and on their skin—the tumbling parasites and symbiotic companions that help make us whole, although Ann tells me it might be possible to revive mammoth bacterial DNA, since, if the tissue is frozen, there's a good chance the tiny piggybacked frozen smidges of bacteria that went with it would be there, too.

So woolly mammoths and golden toads and baiji dolphins and North American camels might all haunt the Earth again. Or perhaps less controversially, the cells might be used to insert more genetic variety into dwindling populations of almost-extinct animals. Saving animals on the brink by diversifying their genome doesn't bother most people; it's a far cry from reincarnating dead ones.

I'm intrigued by the idea of resurrecting Neanderthals, having learned recently that we *Homo sapiens* harbor between 1 and 3 percent Neanderthal DNA, and when I ask Ann her thoughts on the matter, she's clearly fascinated by the mystery.

"I hope I'm a very large amount of Neanderthal!" she says, eyes sparkling. "I mean, how fascinating is it? And it appears to be all in the white Anglo-Saxon gene pool! Maybe *we're* the thick ones. I worry a bit about those guys. Did we polish them all off—or did they die in the cold, or something? I mean if we polished them all off—that's terrible."

Her enthusiasm is refreshing, and I must say, she's the first person I've met who longs to be part Neanderthal, though one of her countrymen, William Golding, wrote a poignantly picturesque novel, *The Inheritors*, narrated from the perspective of the last Neanderthal as his species was being exterminated by our quick, sly, talkative *Homo sapiens* progenitors.

"And in the case of some wonderful extinct animal . . ."

"Saber-toothed tiger?" she offers with clear relish.

"Yes! But would there still be a habitat for saber-toothed tigers?"

"Well, that's debatable. But I personally think biodiversity is good—and if we could bring back some species that we have made extinct, I think I'd rather see them gamboling about in the fields of Kent, or in a woods, than not at all."

Although that's as far as she'll go, Ann realizes that it's something the Ark will need to think about. They could resurrect the woolly mammoth or saber-toothed tiger or dodo or anything else that's extinct. But they've decided, for the moment, to stop when they collect the genetic material. With the DNA of two million to seven million species still to back up, their project is far from finished.

"But it won't stop there," she says with a sibylline smile.

As we well know (think of the cannibal snails on Moorea), introducing new species can have unexpected consequences. And, yes, de-extinction is a divisive topic, even among the diverse members of the Frozen Ark's own consortium. A chief concern from naysayers is that it would divert attention from the serious work of conservation—protecting animals and ecosystems from going extinct in the first place. Critics also worry about the DNA of extinct species weaving through wild populations as ancient newcomers, a different kind of invasive species, one from the past. Both concerns are undoubtedly valid. I'm also troubled by how de-extinction plays into an increasingly mercantile view of life in which most anything is disposable and replaceable by a newer synthetic model. On the other hand, like Ann, I'd really love to see a formerly extinct zebra-like creature "gamboling in the fields of Kent."

Collecting DNA is one thing; agreeing on what to do with it is another. From the Frozen Ark's point of view, there is enough work saving the DNA. Future generations can decide what to do with it in light of the new technologies that emerge. What began as an effort to bank the DNA of only the most endangered animals has now evolved into an urgent banking of whole ecosystems. The Ark goes into an area and collects everything that crawls, flies, scampers, or slithers. In a tropical rainforest with its thick canopy, groups of people spread sheets underneath a tree, and they *shake* it. As I picture raining insects, frogs, snails, and moths, I feel sure Ann finds the shaking and collecting great fun. I know I would. We haven't named more than about 65 percent of the biomass of all the species on Earth. So, yes, *shake* it down, and freeze it, and take it to the Natural History Museum, and label it—so we can tell whether it's an ant or bee or moth—and let the taxonomists name it officially later.

Nottingham stores the DNA of the courtship-crooning Mississippi alligator, giant squid with dinner-plate-sized eye, secretive snow leopard, blue-throated macaw from the Bolivian rainforest, iconic African lion, and square-lipped northern white rhino, among many others. My mind's eye automatically pictures each species in turn. Only about 20 percent of the species are on the endangered species list, and some are not endangered at all. Ideally, the Frozen Ark would store DNA from every species on Earth, but that's not practical. The mammals would be easy, but the bugs would take a long while, especially the beetles, since there are more beetles than anything else on planet Earth (one of my favorites being the dung beetles who navigate by the stars like ancient mariners).

As I set down my cell phone, Ann notices my screen saver of an insanely cute baby wombat, a face that could melt a thousand hearts, and her eyes widen in appreciation. She's just returned from a meeting in Sydney, so I ask her how the cancer-plagued Tasmanian devils are doing.

"They can cure the cancer in captivity," Ann replies, her face showing her concern, "but it's the ones in the wild that are the worry." With a grim nod, she continues, "And the koalas have got chlamydia. And I heard some of the wombats are getting sick."

Just over a hundred of the northern hairy-nosed wombats, the world's cuddliest marsupial, survive in a tiny plot of Queensland. Though once numerous all across Australia, they feed on grasses, and when humans arrived with agriculture and herds of cattle (essentially four-legged mowers), the wombats simply couldn't compete. Drought and invasive species have been polishing off the wombat's supply of native spear, tussock, and poa grasses. It's hard to picture Australia minus most of its famous creatures. Conservationists are trying to treat them in the wild, but Ann doesn't think that Australia's terribly worried about the extinction of its animals.

"They've got this beautiful country and everything looks perfect, and there's not huge numbers of people crashing about," she says.

"So they think it's forever."

"They think it's forever," Ann repeats with incredulous fervor. "But I think Americans are more interested in conservation. Don't you think so?"

In my experience, I tell her, Americans are deeply concerned about conservation, but we have clashing, fiercely defended opinions about how to do it. Some believe it's essential to preserve our majestic national parks; some, that the parks are a lost crusade and that safeguarding animals in big preserves just hasn't worked. Some believe in rewilding's networks of "cores, corridors, and carnivores" to reconnect and rebalance unstable ecosystems; or Pleistocene rewilding—in North America, unloosing elephants, lions, bison, and cheetahs (the closest living relatives of the ancient native megafauna) to roam the Great Plains. Others argue that all of the above are last-epoch thinking, and, as an increasingly metropolitan species, we should weave more of the wild into the cities where we live. Ann feels certain that we need multifaceted solutions.

"You've got to try the wild, but obviously that's all going to go west if one's being realistic."

For a moment, I think perhaps she means the expanse of America's West, then quickly realize she means die, as in following the setting sun.

Ann says firmly, "You've got to have parks. And you've got to manage your wild."

"Managed wilderness. You don't think we can afford to just let nature run wild anymore?"

"No," she says with conviction. "We really can't."

Nature has become too fragmented to just run wild.

Ann tells me of a local solution that works: how in English towns, where terraced housing is commonplace, small back-to-back walled gardens lead onto each other and combine to create long fertile corridors for wildlife.

"So when I go into Cambridge and do a bit of gardening," she says, "I'm surrounded by all sorts of insects and mosses and butterflies. But in our country home's garden, there's nothing. Consider starlings. They put their beaks down about four centimeters into the soil—that's how they feed—and that's where all the pesticides are accumulating in the agricultural areas. The environment would be absolutely fine for them *if* there just weren't any pesticides in it!"

Hoping to lure pollinators back, what's known as Plan Bee rewards landowners for planting wildflower meadows for knapweed, bird's-foot trefoil, red clover, and other weeds favored by insects, butterflies, and bees. They're called "bee roads," perhaps in keeping with the Anglo-Saxon tradition of referring to the ocean as the "whale road." Other English bees have become prosperous city-dwellers, unassailed by agricultural pesticides, and there are now more sparrows, starlings, and blackbirds in the town gardens than in the open countryside.

A sorry image of the English countryside, silent at dawn and devoid of wings, slinks through my mind. All the more reason I like Ann's "try everything" mindset. Yes to national parks, to rewilding preserves, to wildlife corridors, to city shades of green, to DNA banking, and to any other strategy we can think of that will allow animals to pursue their dusty, feral ways and nature to stay replete with potent life forms.

The office door swings gently open, and Chris Wade pokes his head in to take me on a tour of the lab. A tall man with dark curly hair, Chris does DNA analysis on the samples that arrive by mail, and he's looking forward to being part of the Ark's upcoming expedition to Vietnam to collect fresh DNA samples. Walking across the outer office, we enter a shared college lab with pale-blue walls, workbenches, microscopes, and a bevy of white lab coats hanging on one wall like a colony of albino bats.

Chris explains that they take the DNA field samples in several forms to be sure of backups. Dropping a tissue slice into a tube and topping it up with ethanol essentially pickles the tissue. That state of DNA isn't ideal; it's not as high-quality as fresh-frozen DNA. But if they put a specimen in a freezer, and the freezer were to break—and in a lot of countries that's often a problem—then it would be completely destroyed.

You'd think freezing would kill the cells, but the arkmasters control the rate of cooling, at only a degree a minute, and that slow-motion plunge keeps the cells intact. For safety, they prefer a three-way approach:

"One: Ethanol. Tough as nails," Chris says, "but there for you in the end—the freezer can go off, everything can go wrong, you will still have that preserved bit. Two: We've also got our fresh-frozen tissue slice, which is *perfect*, it's got everything. Three: And then we've got another sample for later cell culture. That's the ideal scenario, all three methods."

He leads me to the far end of the lab, through a doorway painted cornflower blue, into a small room, where four tall white Sanyo freezers stand, looking surprisingly humdrum, suitable perhaps for a wintering farm family, not the biggest snow survival fort the world has ever known. I put on green latex gloves to protect my primate skin.

He tugs opens a freezer door that gusts a small white cloud, like the combined souls of ten thousand animals exhaling in unison on an ice floe. Inside the freezer sit row after row of frost-covered drawers. Pulling one open with a gloved hand, I'm surprised to find only carefully arranged rows of short frost-covered vials, each with a label.

"You can lift one out," Chris says in a tone of voice warning *But carefully!*

When I do, I see that I'm holding the future of an African lion. If I blink, a tawny-colored male lion with a shaggy mane is standing on the veldt of my palm. I blink again and the whole animal is nestled in my hand, with room to spare for tall grass, heat mirage, and his pride. What an unlikely way to safeguard the future of animal-kind.

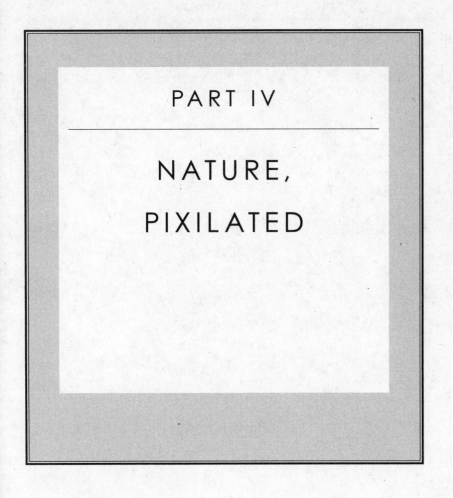

PART IV

NATURE,
PIXILATED

AN (UN)NATURAL
FUTURE OF THE SENSES

What about us? Are *we* natural anymore? How can we be, when we've morphed into superheroes? Our ancestors adapted to nature according to the limits of their senses. But over the eons, by extending our senses through clever inventions—language, writing, books, tools, telescopes, telephones, eyeglasses, cars, planes, rocket ships—we've changed how we engage the world and also how we think of ourselves. We just assume now that human beings can move across the skies at 500 mph. Or spot a hawk across a valley. Or do colossal calculations at speed. Or watch events unfolding halfway around the world. Or safely repair someone's heart. Or wage war. Our attitude about our own nature, what sort of creatures we are, now includes the novelties we've pinned to our senses.

All these add-ons are a perfectly ordinary part of daily life. The use of tools and technology has become an innate part of our being, as we extend ourselves deeper and deeper into our environment. In the past decades a fundamental change has evolved in the idea of the universe we inhabit, and also what a human being is and may become. We don't worry if we can't see a splinter in a child's finger. We automatically don glasses and become an animal with keener eyesight. That may save the child from infection, but it also revises what a human being is. How will that continue changing in our lifetime?

Already, we're masters of the invisible. Just as we accept that the universe is mainly invisible dark matter and dark energy, we accept the reality of protozoans and viruses even though we can't see them without a microscope, or perhaps as stationary oddities in the pages of a textbook—which few people are tempted to do. We believe in television and radio waves, gnomelike quarks, GPS, microwaves, the World Wide Web, gosling photons, a mantilla of nerve endings in the brain, the voiceless hissing of background fizz from the Big Bang, planets orbiting many stars in the night sky—some hospitable to life. Then there's all the panting eyes, throbbing jellies, iridescent bladders, and glowing mouths haunting the remote sunless abysses of the deep sea.

Our mental cosmos teems with a thicker texture of invisibles than ever before. Living with invisible forces used to mean spirits, ghosts, gods, angels, and ancestors. Our view of nature now supplies different familiar ghosts, including all the wispy tangles, tinctures, and driblets of a working body being revealed to us as never before through technology and nanotechnology. We take for granted the vast invisible worlds surrounding and inside of us. It's a sort of high-tech shamanism (the belief that spirits inhabit all things, living or nonliving). Some entities may hide in the holly bush at the front door; others float light-years away.

We can forge so many invisibles in our mind's eye because enough of our kind have witnessed them firsthand, through microscopes, telescopes, or computers, and smeared that knowledge far and wide. As a result, the air gyrates with invisibles I can hear but not see, and yet take for granted like distant relatives whose photos I've framed.

In autumn, a season of night fiddlers, I know summer is fraying away because the air brims with their eerie music, although I don't see the hidden musicians—katydids and crickets playing their marimbas, as they lift their wings high and rub a sharp edge of one wing over a ridge of pegs on the other. It's not easy to spot cellophane-winged aerobats among late summer's wild chicory, Queen Anne's lace, and clover-scented milkweed—kingdom of the giant, much showier monarch butterfly.

Nonetheless, I can picture them all combing strands of song from their wings, picture them in microscopic hair-perfect detail. Katy-dids are rasping a tattletale: *Katy did! Katy did! Katy did!!* Cicadas, buckling and unbuckling their stomach muscles, are yielding the sound of someone sharpening scissors. Fall field crickets, the ther-mometer hounds, are adding high-pitched tinkling chirps to the jazz. Carolina crickets are furnishing a buzzing trill. Grasshoppers sound like they're shuffling decks of cards. Snowy tree crickets are lending an evenly spaced chirping melody to the ensemble. It's the ultimate jug band, using body parts as instruments.

I don't see any of their courtship, since they're small and hidden in the darkness. But I've learned enough from scientist-seers and their technology to trust that the males do all the serenading, horny for females, each of whom waits in the dark loins of the night, listen-ing with ears in rather odd places—on the abdomen or the knees. She homes in on a winged dude, lured by his siren song. Then the happy male croons a different courtship tune. But they haven't much time for dalliance before the first heart-stopping frost. According to folklore's timetable—and I still believe in folklore—frost creeps in ninety days past the katydids' first song. In my insect-loud yard, I heard the first katydid call about a week early this year, round about the middle of July, and sure enough frost fell in mid-October.

Alongside this buzzing-chirping-tinkling-fiddling in the night, and choreographed to it, there's the raw sexcapade drama. And, although I don't ogle thousands of bugs *in flagrante* all over the woods, and tens of thousands, maybe millions, yodeling their lust downtown, up in the forest leaf-parlors, and along sinewy country roads, I was a college student once; I get it.

All of this happens unseen, which is a haunting thought, but even without laying our eyes on the crickets, grasshoppers, cicadas, and katydids, we hear them shrilling in this season and trust that they're the tiny living gargoyles scientists claim. We believe the katydids exist in their scratchy little corner of the invisible—an act of faith that suits us just fine. Most don't wish to search in the dark buzz of night for the multieyed and antenna-ed.

Anyway, these days, we know we can verify the existence of creatures we can't see easily enough in books, films, or bug Facetime on the Internet. The ancients believed the gods were angry when storms crackled and boomed. We check the Weather Channel's radarscope.

Our ancient understanding of nature (faith, lore, hearsay, story) has a new level, one changing almost every day, proxy *sine nomina* from technology-equipped scientists and other researchers, the designated witnesses who behold, listen, and chronicle as the likes of insect love parties on. We agree en masse to believe these professionally designated seers.

Or we become citizen-seers ourselves. A smartphone will do. Walking on a trail in New Forest, in southern England, I stopped in a sunlit clearing when I heard the distinctive rasp of Buddhist monks creating a sand mandala by rubbing brass scrapers down ribbed brass funnels to release single brightly colored grains. A quick spin around: trees, pastures, a perching woodland warbler falling into flight, shadows dancing along the trail. No monks. I smiled at the sleight of ear. Weeks before, at Keystone College in Pennsylvania, I had watched, enchanted, as Tibetan monks sand-dribbled a mandala, producing what sounded like the trills and quiverings of invisible cicadas.

Pulling my iPhone out of my pocket, I opened the Cicada Hunt app, a brainchild of entomologists at the University of Southampton. Over a thousand people sent in reports this past summer. On the iPhone, a green card appeared with a white cicada icon sitting on black velvet at its center. When I held it up and tapped the cicada icon, a white outer ring fanned open around it and the cicada glowed orange. For eighteen seconds, the app tested the soundscape for the exact frequency of rare New Forest cicadas. No luck. Only a scant few New Forest cicadas have been detected by the thousands of citizen scientists in England since 2000, but that's enough to offer proof of their whereabouts and need for protection. Though I knew it was a long shot, I found the app doesn't register the calls of my homely New York variety of cicada.

The Buddhist mandala-makers may live in a cosmos dancing with colorful deities, just as they always have. But now they and the Dalai Lama (a science aficionado) are also aware, from mindful moment to moment, of an invisible dimension that includes neurons, quarks, Higgs bosons, MRIs, condensation nuclei, white dwarfs, DNA, and a googolplex of others.

Elsewhere on Earth, over 5.2 million Internet-connected computers, citizen scientists are helping SETI (Search for Extraterrestrial Intelligence) monitor radio telescope data through the SETI@home project, hoping to catch a message from alien life forms in some distant star system. SETI's senior astronomer, Seth Shostak, believes that the first calling card from aliens may well be detected on home computers, not by official scientists at radio telescopes arrayed in India, Australia, Puerto Rico, or Chile.

More than ever, our technology allows us to peer into worlds far beyond our outmoded senses, into a realm where cells loom large as lakes, pores are chasms, the body is just another kind of ecosystem, and the idea of cartography no longer applies only to landforms. We've mapped galaxies and genomes. We keep projecting ourselves into landscapes we're not equipped to cross in the flesh. Computers have shed light on biological processes invisible to humankind until very recently. In 1990, I wrote about our sensory grasp of the world in *A Natural History of the Senses*. Only twenty years later, the basic experience is the same, but its scope has been vastly amplified. For example, our proprioception, the sense of where we are in space, now spins far beyond the physical body. We can spy on ourselves in sly, public, or cloak-and-dagger ways, from lavish perspectives, inside and out. By satellite, a drone's eye, via Skype, on security cameras, through electron microscopes. Some of us are even relaxed about, or excited by, the promise of connecting our brains to the world outside of the body. In such sweeping sensory adventures, our cameo of who and what we are shifts, and also how we may decide to know ourselves in the future.

What we see and think when we look at the night sky has also changed. Two decades ago, the only planets were here in our own solar system. Now we know that the cosmos is littered with them. We know now that the Milky Way, the backbone of night, is twice as large, even heavier, and spinning faster than we previously thought. Also that it has four arms, not two. Our telescopes listen with cupped ears for whispers from the beginning of time, when the whole universe was no larger than a grapefruit, a small solid object, before the light of stars and the destiny of planets. How could something that small give birth to more space than the mind's eye can fathom?

Although the brain's star chamber is sealed and invisible in its cave of bone, we're craning our high-tech senses (MRI, fMRI, PET scan, etc.) to peer in as never before at networks lit like night views of Earth from space. Thanks to digital displays, scalpel-less dissecting of live patients is commonplace, as is cut-free slicing of gray matter into wafer-thin sheets that can be viewed three-dimensionally and rotated as if the conscious, alert, and no doubt mind-wandering occupant had set his actual brain on an anatomy bench for anyone to probe. All sorts of abnormalities and diseases, such as schizophrenia and autism, have bared some of their bones, and we've begun exploring the mental haunts of such notorious intangibles as religion, addiction, and compassion. By studying busy neural work sites, increased traffic flow, and where thought-crews guzzle oxygen as they toil, we're forming insights about everything from lying to love. For the first time, we're able to see some of the ties that bind us. The verb we use, "scan," which used to mean a brief skim with the eyes, has evolved into its opposite: a machine's searching stare. People gamely volunteer to have their heads examined so that researchers can witness emotional regattas in full sail (or, sometimes, on the rocks).

Nightly news often reports the latest nugget about concussion, depression, rejection, multitasking, empathy, risk-taking, fear, and other states of mind—explained in terms of the neural architecture and wiring of the brain. In 2012, when President Obama proposed a federally funded $100 million brain-mapping project, stressing that "as humans we can identify galaxies light-years away, study particles smaller than an atom, but we still haven't unlocked the mystery of the three pounds of matter that sits between our ears," some people balked at the expense, but few believed it wasn't possible and a worthy goal.

A new field, called interpersonal neurobiology, draws its vigor from one of the great discoveries of our age: the brain is rewiring itself daily, and all relationships change the brain—but especially our most intimate bonds, which foster or fail us, altering the delicate circuits that shape memories, emotions, and that ultimate souvenir, the self. Love is the best school, but the tuition is high, and the homework can be physically painful. As imaging studies by the UCLA neuroscientist Naomi Eisenberger show, the same areas of the brain that register physical pain are active when someone feels brutalized by love. That's why rejection hurts all over the body, but in no place you can point to. Or rather, you'd need to point to the dorsal anterior cingulate in the brain, the front of a collar wrapped around the corpus callosum, the bundle of nerve fibers zinging messages between the hemispheres, that registers both rejection and physical assault. Whether they speak Armenian or Mandarin, people around the world use the same images of physical pain to describe a broken heart, which they perceive as *crushing, crippling, a real blow* that *hurts so bad* they *go all to pieces*. It's not just a metaphor for an emotional punch that's too shadowy to name. As our technology is beginning to reveal, social pain—rejection, the end of an affair, bereavement—can trigger the same sort of sensations as a stomachache or a broken bone.

But a loving touch is enough to change everything. The neuro-scientist James Coan, of the University of Virginia in Charlottesville, conducted experiments in which he gave an electric shock to one ankle of women in happy committed relationships. Tests registered their anxiety before and pain level during the shocks. Then they were shocked again, this time holding their loving partner's hand. The same level of electricity produced significantly lower pain and even less neural response in the cingulate. In troubled relationships, this protective effect did not occur. If you're in a healthy relationship, holding your partner's hand is enough to subdue your blood pressure, ease your response to stress, improve your health, and even soften physical pain. We're able to dramatically alter one another's physiology and neural functions—and watch.

The ability to see these scans has ushered in a whole new level of relating to one another. One can decide to be a more attentive and compassionate partner, mindful of the other's motives, hurts, and longings. Breaking old habits and patterns isn't easy, but couples are choosing to rewire their brains on purpose, sometimes with a therapist's help, to ease conflicts and strengthen their at-one-ness. Neanderthals didn't sit around thinking about their partners' neurons—and neither did Plato, Shakespeare, Michelangelo, or my mother, for that matter. I didn't when I was an undergraduate. Even though we are still in the early days of brain imagery, we're tagging invisibles like butterflies; we're learning life-altering truths.

What will this mean for a new Anthropocene ethics? How might our knowledge influence how we choose to relate to our spouse, children, friends, coworkers? As such knowledge trickles through society, will it influence how we conduct our relationships? How will we handle the responsibility of knowing that harsh words can be as physical as a punch, inflict violent pain, and subtly mess with the wiring in someone's brain?

WEIGHING IN THE NANOSCALE

We're not just seeing invisibles; we're engineering things on a minute, invisible-to-the-eye scale. "Nano," which means "dwarf" in Greek, applies to things one-billionth of a meter long. In nature that's the size of sea spray and smoke. An ant is about 1 million nanoparticles long. A strand of hair is 80,000 to 100,000 nanometers wide, roomy enough to hold 100,000 perfectly machined carbon nanotubes (which are 50 to 100 times stronger than steel at one-sixth the weight). A human fingernail grows about 1 nanometer a second. About 500,000 nanometers would fit in the period at the end of this sentence, with room left over for a rave of microbes and a dictator's heart.

I'm stirred by the cathedral-like architecture of the nanoscale, which I love to ogle in photographs taken through scanning electron microscopes. One year in college, I spent off-duty hours hooking long-stranded wool rugs after the patterns of the amino acid leucine (seen by polarized light), an infant's brain cells, a single neuron, and other objects revealed by such microdelving. How beautifully some amino acids shine when lit by polarized light: pastel crystals of pyramidal calm, tiny tents along life's midway. Arranged on a slide or flattened on a page, they glow gemlike but arid. We cannot see their vitality, how they collide and collude as they build behavior. But their nanoscale physiques are eye-openers, and more and more we're turning to nature for inspiration.

We used to think that wall-climbing geckos must have suckers on the soles of their feet. But in 2002, biologists at Lewis & Clark College in Portland, Orgeon, and the University of California at Berkeley released their strange findings, and science was agog. Viewed at the nano level, a gecko's five-toed feet are covered in a series of ridges, the ridges are tufted with billions of tiny tubular elastic hairs, and the hairs bear even tinier spatula-shaped boots. The natural force between atoms and molecules is enough to stick the spatulas to the surface of most anything. And the toes are self-cleaning. As a gecko relaxes a toe and begins to step, the dirt slides off and the gecko steps out of it. No grooming required.

When I learned of gecko feet from a biologist friend with an infectious sense of wonder, the idea of *sticky* instantly changed from a gluey sensation to a triumph of nature's engineering. The next time I spied a gecko climbing up a stucco wall, my brain *saw* the tidy toes rising, and the spatula-tipped hairs clinging, even though my raw eyes couldn't see beyond the harlequin slither. Inspired by gecko toes, scientists have invented chemical-free dry bio-adhesives and -bandages, and all sorts of biodegradable glues and geckolike coatings for home, office, military, and sports.

The nanotechnology world is a wonderland of surfaces unimaginably small, full of weird properties, and invisible to the naked eye, where we're nonetheless reinventing industry and manufacturing in giddy new ways. Nano can be simply, affordably lifesaving during natural disasters. The 2012 spate of floods in Thailand inspired scientists to whisk silver nanoparticles into a solar-powered water filtration system that can be mounted on a small boat to purify water for drinking from the turbid river it floats on.

In the Namibian desert, inspired by water-condensing bumps on the backs of local beetles, a new breed of water bottle harvests water from the air and refills itself. The bottles will hit the market in 2014, for use by both marathon runners and people in third-world countries where fresh water may be scarce. South African scientists have created water-purifying tea bags. Nano can be as humdrum as the titanium dioxide particles that thicken and whiten Betty Crocker frosting and Jell-O pudding. It can be creepy: pets genetically engineered with firefly or jellyfish protein so that they glow in the dark (fluorescent green cats, mice, fish, monkeys, and dogs have already been created). It can be omnipresent and practical: the army's newly invented self-cleaning clothes. It can be unexpected, as microchips embedded in Indian snake charmers' cobras so that they can be identified if they stray into the New Delhi crowds. Or it can dazzle and fill us with hope, as in medicine, where it promises nano-windfalls.

In the 1966 science-fiction movie *Fantastic Voyage*, a tiny human-crewed submarine could sail through a patient's turbulent bloodstream, careening down the rapids of an artery, dodging red blood cells, drifting through flesh lagoons, until they found the diseased or torn parts needing repair. With the advent of nanotechnology, this adventure leaves the realm of fiction. Researchers are perfecting microscopic devices known as nanobots and beebots (equipped with tiny stingers) that can swim through the bloodstream and directly target the site of a tumor or disease, providing radical new treatments.

The futurist Ray Kurzweil predicts that "by the 2030s we'll be putting millions of nanobots inside our bodies to augment our immune system, to basically wipe out disease. One scientist cured Type I diabetes in rats with a blood-cell-size device already."

There are nanobots invisible to the immune system, which shed their camouflage when they reach their work site. Tiny and agile enough to navigate a labyrinth of fragile blood vessels, some are thinner than a human hair. Researchers at the École Polytechnique de Montréal in Canada are developing a kind of self-propelled bacterium with naturally magnetic innards. In nature, the bacterium's corkscrewlike tail propels it, and its magnetic particles point like a compass needle to guide it toward deeper water and away from the death knell of oxygen. Researchers are learning to steer the bacterium with precise tugs and pushes from an MRI machine, and at only 2 microns in diameter, the bacteria are small enough to fit through the smallest blood vessels in the human body. These harnessed bacteria can carry polymer beads roughly 150 nanometers in size; the goal is to modify the beads to carry medicines to tumors and other targets. Because we find it hard to imagine both ends of the visual spectrum—the cosmic infinite or the minutely finite—it sounds impossible. Yet we believe in them as surely as we do the unseen katydids in the woods.

We may imagine harnesses as large, leathery, and worn, and bacteria as invisibly tiny, able to slip into, through, and around objects or people. Harnessing bacteria doesn't form a feasible image in the mind's eye. You need to imagine bacterial horses and magnetic harnesses carrying polymer-bead bells that jingle a cancer-fighting drug. Still, a "sleigh" of medicine could become a new commonplace that slips into conversation the way a "flight" of stairs has, so comfortably that we no longer picture birds in flight when we see a staircase. We're constantly minting new metaphors for the brain to use as mental shortcuts.

"Is there a sleigh for my illness?" someone may one day ask a doctor, as we now ask, "Is there a pill I can take?"

Because boys love monster machines that dig, drag, roar, or explode, maybe the metaphor will be a "tug," "tractor," "missile," or "submarine." It might even be a "Phelps," after the Olympic swimmer.

Another recent marvel of nanotechnology promises to alter daily life, too, but this one, despite its silver lining, is wickedly dangerous. Inevitably, it will inspire a welter of patents and ignite bioethical debates. Nano-engineers have devised a true silver bullet, a way to coat both hard surfaces (such as hospital bedrails, doorknobs, and furniture) and also soft surfaces (sheets, gowns, and curtains) with microscopic nanoparticles of silver, an element known to kill microbes. You'd think the new nanocoating would be a godsend to patients stricken with hospital-acquired sepsis and pneumonia and to the doctors fighting what has become a nightmare of antibiotic-resistant microorganisms that kill forty-eight thousand people a year.

It is. That's the problem.

It's possibly too effective. Remember, most microorganisms are harmless, many are beneficial, but some are absolutely essential for the environment and human life. Bacteria were the first life forms on the planet, and we owe them everything. Swarms of bacteria blanket us, other swarms colonize our insides, and still more flock like birds to any crease, cave, or canyon of the body they can find. Our biochemistry is interwoven with theirs. We also draft bacteria for many industrial and culinary purposes, from decontaminating sewage to creating tangily delicious foods like kefir, sauerkraut, and yogurt. So we need to be careful about the bacteria we target.

Will it be too tempting for nanotechnology companies, capital-izing on our fears and fetishes, to engineer superbly effective nano-silver microbe-killers, deodorants, and sanitizers of all sorts for home and industry? We may accept the changes nanotechnology creates in everyday life (such as antimalaria garments that ward off bugs) as part of the brave new world we deserve, yet we're inventing them before thinking through their potential consequences. There's no evidence that the antibacterial soaps available at the supermarket work better than soap and water, and in fact they may be hazardous. Triclosan—one of the standard ingredients in these soaps—is con-sidered a pesticide by the FDA.

That's why Kathleen Eggleson, a scientist at the University of Notre Dame, founded the Nano Impacts Intellectual Community, a monthly meeting that draws campus researchers, community lead-ers, and visiting scholars and authors to discuss the ethics and impact of new developments in nanotechnology. Her April 2012 paper pub-lished by the Center for Nano Science and Technology highlights the risk of unregulated products destroying microbial biodiversity. "Just this past December," she points out, a coating for textiles "was the first nano-scale material approved as a pesticide by the FDA." What if our nanopesticides accidentally kill the nitrogen-fixing bacteria that make our atmosphere breathable?

How incredible that we now have national committees and college seminars that debate bioethics, neuroethics, and nanoethics. We're creating ethical predicaments that would have made Montaigne or Whitman blink. "I sing the body electric," Walt Whitman wrote in 1855, inspired by the novelty of useful (not just parlor-trick) electricity, which he would live to see power streetlights and telephones, trolleys and dynamos. Whitman was the first American poet the technological universe didn't scare. He often celebrated the steam engine, the railroad, and other new inventions of his era. In *Leaves of Grass*, his ecstatic epic poem of American life, he depicts himself as a live wire, a relay station for all the voices of Earth, natural or invented, human or mineral. "I have instant conductors all over me," he wrote. "They seize every object and lead it harmlessly through me. . . . My flesh and blood playing out lightning, to strike what is hardly different from me."

The invention of electricity equipped Whitman and other poets with a scintillation of metaphors. Like inspiration, it was a lightning flash. Like prophetic insight, it illuminated the darkness. Like sex, it tingled the flesh. Like life, it energized raw matter. Deeply as he believed the vow "I sing the body electric," Whitman didn't know that our cells really do generate electricity, that the heart's pacemaker relies on such signals, and that billions of axons in the brain create their own electrical charge (equivalent to about a 60-watt bulb). A force of nature himself, he admired the range and raw power of electricity.

Yet I'm quite sure nanotechnology's recent breakthroughs would have stunned him, such as the dream textile named GraphExeter, a light, supple, diaphanous material made for conducting electricity, which will revolutionize electronics by making it fashionable to wear your computer or cell phone. Recharging would be automatic as nanosized generators convert the body's normal stretches and twists into electricity through the piezoelectrical effect (what keeps a self-winding quartz watch ticking). Wake Forest engineers recently invented Power Felt, a nanotube fabric that generates electricity from the difference in temperature between room heat and body heat. You could start your laptop by plugging it into your jeans, recharge your cell phone by tucking it into a pocket. Then, not only would your cells sizzle with electricity, even your couture clothing could chime in.

Would a fully electric suit upset flight electronics, pacemakers, airport security monitors, or the brain's cellular dispatches? If you wore an electric coat in a lightning storm, would the hairs on the back of your neck stand up? Would you be more prey to a lightning strike? How long will it be before late-night hosts riff about electric undies? Will people tethered to recharging poles haunt airport waiting rooms? Will it become hip to wear flashing neon ads, quotes, and designs—maybe a lover's name in a luminous tattoo?

Yet electricity has already lost its pizzazz. It's hard to spot things hidden in plain sight. Even harder when they're invisible. We take electricity for granted, unaware of it if lights and devices are turned off. Still, its specter haunts the walls all around us, sizzles in great looping strings that encircle us. If you have sockets in your house, you keep pocket lightning pulsingly at hand. Flip one switch and daylight floods the room; flip another and night falls like an iron door. The ancient Romans used to build their spas around natural hot springs; today we keep miniature electric hot springs in our homes to boil water for washing and bathing. Electric clocks watch over us while we sleep, an electric furnace (even gas or oil heat uses an electric pilot light) keeps us warm, and an electric fan or air conditioner cools us. In the summer we live in an electric igloo.

How Anthropocene that we "condition" the very air we breathe, flavoring its essence. For most of human history, we simply breathed the air that surrounded us, whatever nature delivered, whether it was fume-laced from oil deposits or salty-fishy from the coast. Before the Industrial Revolution, neighbors inhaled similar air. Now we tailor the breath that streams into and out of us. Neighbors may prefer their homes warmer, cooler, candle-scented, more humid, redolent of ammonia or bleach, cleaned by UV lights or "ozone-spiced" bulbs. We personalize our air!

We don't find any of this strange, don't regard it as unnatural. We don't even notice it unless the electricity goes out, and then it's as if the electric in our cells failed and we feel *disconnected*, a word we use when speaking of both power outages and psychic alienation (which feels like our inner grid has blown a million fuses). We too are electric, after all, a hive of minute, usually imperceptible jolts, as electrical signals leap like mountain goats from cell to cell throughout the body. Electricity, the brain's telegraph, is almost instantaneous. Pinch a nerve in your back and dancing knives prick your skin. Pinch a nerve in your neck and a tiny electrocutioner throws a switch. But sexual tingles and jolts we find shockingly pleasurable, "electric flesh-arrows," as Anaïs Nin calls them.

Electricity is a molecular tug-of-war. Life forms can't exist without electric pumps in each cell. Ions of potassium and sodium, flowing into and out of a cell, produce a wave. The sodium is forced out, the potassium rushes in, the potassium is pushed out of another cell, the sodium rushes in, and so on. Ions fly like balls tossed by a one-armed juggler. It's balance gone awry, regained, and lost again.

We reject things deemed too "wobbly," "rickety," or "unsteady." We may condemn a person for being "unstable," "unhinged," or "unbalanced." Yet deep in every cell, even the most slothful of us are falling out of balance and recovering. The body's inner electric is not a steady stream. How ironic it is that we fight change in our lives and yearn for a state of permanence no life form can manage without dying—because we're forever tumbling and snatching ourselves up before we smack the dirt and stay down.

Just as electricity ghosts throughout modern buildings everywhere and all the time, the same will soon be true of digital technology, woven into the walls, flowing through the floors, hidden all around and upon us. We'll completely clothe ourselves in it, swim in it. As with the natural and man-made electric in our lives, we'll probably ignore the clouds of technology we float on, under, and inside everywhere all the time. The brain relishes familiarity, loves being on autopilot, because then it can slur over the details and spend most of its sparking on something else. At the Oshkosh Airshow I attended one summer, the first wing-walker atop an old biplane drew gasps from thousands of people brought to a halt in amazement. The second fetched an anthology of admiring stares. By the third, a surprising number of people were blasé and continued milling about, chatting or shopping. You could almost hear their brains moaning, *Oh, that again, another wing-walker.* It doesn't take much for a novelty to become invisible. And yet, isn't this what we wish from exciting new technology, for it to slide invisibly into our lives, making them effortless and more enjoyable?

We'll get used to living inside a digital bubble. Unless, perhaps, we must think hard when we connect via a brain-computer interface to the house's fixtures. But even then, habit being what it is, we'd most likely come home, absentmindedly recall the code opening the front door's lock, step over the threshold, daydream a hand swiping a light switch until the room brightens, mind's-eye visualize the solar-electric shingles melting ice jams on the roof, while simultaneously worrying over a supposed affront at work or rebuff at school, anticipating dinner, fantasizing about a cute guy or gal, and hearing a stupid tune lodged in some spiky thicket of the brain.

Whether it's hospital chairs robed in silver nanojackets to ward off bacteria, or invisibility cloaks, or degradable electronic devices that dissolve when you're finished with them, or thin, flexible solar panels that can be printed or painted onto a surface, the writing is on the wall (though you'll need a microscope to read it). And when it comes to the delicate balance of Earth's life forms, it may be a small, small world after all.

NATURE, PIXILATED

t is winter in upstate New York, on a morning so cold the ground squeaks loudly underfoot as sharp-finned ice crystals rub together. The trees look like gloved hands, fingers frozen open. Something lurches from side to side up the trunk of an old sycamore—a nuthatch climbing in zigzags, on the prowl for hibernating insects. A crow veers overhead, then lands. As snow flurries begin, it leaps into the air, wings aslant, catching the flakes to drink. Or maybe just for fun, since crows can be mighty playful.

Another life form curves into sight down the street: a girl laughing down at her gloveless fingers, which are busily texting on some handheld device. This sight is so common that it no longer surprises me, though strolling in a large park one day I was startled by how many people were walking without looking up, or walking in a myopic daze while talking on their "cells," as we say in shorthand, as if spoken words were paddling through the body from one saltwater lagoon to another.

We don't find it strange that, in the Human Age, slimy, hairy, oozing, thorny, smelly, seed-crackling, pollen-strewn nature is digital. It's finger-swiped across, shared with others over, and honeycombed in our devices. For the first time in human history, we're mainly experiencing nature through intermediary technology that, paradoxically, provides more detail while also flattening the sensory experience. Because we have riotously visual, novelty-loving brains, we're entranced by electronic media's caged hallucinations. Over time, can that affect the hemispheric balance of the brain and dramatically change us? Are we able to influence our evolution through the objects we dream up and rely on?

We may possess the same brain our prehistoric ancestors did, but we're deploying it in different ways, rewiring it to meet twenty-first-century demands. The Neanderthals didn't have the same mental real estate that modern humans enjoy, gained from a host of skills and preoccupations—wielding laser scalpels, joyriding in cars, navigating the digital seas of computers, iPhones, and iPads. Generation by generation, our brains have been evolving new networks, new ways of wiring and firing, favoring some behaviors and discarding others, as we train ourselves to meet the challenges of a world we keep amplifying, editing, deconstructing, and recreating.

Through lack of practice, our brains have gradually lost their mental maps for how to read hoofprints, choose the perfect flints for arrows, capture and transport fire, tell time by plant and animal clocks, navigate by landmarks and the stars. Our ancestors had a better gift for observing and paying attention than we do. They had to: their lives depended on it. Today, paying attention as if your life depends on it can be a bugbear requiring conscious effort. More and more people are doing all of their reading on screens, and studies find that they're retaining 46 percent less information than when they read printed pages. It's not clear why. Have all the distractions shortened our attention spans? Do the light displays interfere with memory? It's not like watching animals in ordinary life. Onscreen, what we're really seeing isn't the animal at all, but just three hundred thousand tiny phosphorescent dots flickering. A lion on TV doesn't exist until your brain concocts an image, piecemeal, from the pattern of scintillating dots.

College students are testing about 40 percent lower in empathy than their counterparts of twenty or thirty years ago. Is that because social media has replaced face-to-face encounters? We are not the most socially connected we've ever been—that was when we lived in small tribes. In our cells and instincts, we still crave that sense of belonging, and fear being exiles, because for our ancestors living alone in the wild, without the group protection of the tribe, meant almost certain death. Those with a strong social instinct survived to pass their genes along to the next generation. We still follow that instinct by flocking to social media, which connects us to a vast multicultural human tribe—even though it isn't always personal.

Many of our inventions have reinvented us, both physically and mentally. Through texting, a child's brain map of the thumbs grows larger. Our teeth were sharper and stronger before we invented cooking; now, they're blunt and fragile. Even cheap and easily crafted inventions can be powerful catalysts. The novelty of simple leather stirrups advanced warfare, helped to topple empires, and introduced the custom of romantic "courtly" love to the British Isles in the eleventh century. Before stirrups, wielding either a bow and arrow or a javelin, a rider might easily tumble off his horse. Stirrups added lateral stability, and soldiers learned the art of charging with lances at rest, creating terror as their horses drove the lances home. Fighting in this specialized way, an aristocracy of well-armed and -armored warriors emerged, and feudalism arose as a way to finance these knights, whose code of chivalry and courtly love quickly dominated Western society. In 1066, William the Conqueror's army was outnumbered at the Battle of Hastings, but, by using mounted shock warfare, he won England anyway, and introduced a feudal society steeped in stirrups and the romance of courtly love.

Tinkering with plows and harnesses, beyond just alleviating the difficult work of breaking ground, meant farmers could plant a third-season crop of protein-rich beans, which fortified the brain, and some historians believe that this brain boost, right at the end of the Dark Ages, ushered in the Renaissance. Improved ship hulls spread exotic goods and ideas around the continents—as well as vermin and diseases. Electricity allowed us to homestead the night as if it were an invisible country. Remember, Thomas Edison perfected the lightbulb by candle or gas-lamp light.

Our inventions don't just change our minds; they modify our gray and white matter, rewiring the brain and priming it for a different mode of living, problem-solving, and adapting. In the process, a tapestry of new thoughts arises, and one's worldview changes. Think how the nuclear bomb altered warfare, diplomacy, and our debates about morality. Think how television shoved wars and disasters into our living rooms, how cars and airplanes broadened everything from our leisure to our gene pool, how painting evolved when paints became portable, how the printing press remodeled the spread of ideas and the possibility of shared knowledge. Think how Eadweard Muybridge's photographs of things in motion—horses running, humans broad-jumping—awakened our understanding of anatomy and everyday actions.

Or think how the invention of the typewriter transformed the lives of women, great numbers of whom could leave the house with dignity to become secretaries. Although they won the opportunity because their dexterous little fingers were considered better able to push the keys, working in so-called pools they risked such bold ideas as their right to vote. Even the low-tech bicycle modified the lives of women. Straddling a bike was easier if they donned bloomers—large billowy pants that revealed little more than that they had legs— which scandalized society. They had to remove their suffocating "strait-laced" corsets in order to ride. Since that seemed wicked, the idea of "loose" women became synonymous with low morals.

In ancient days, our language areas grew because we found the rumpled currency of language lifesaving, not to mention heady, seductive, and fun. Language became our plumage and claws. The more talkative among us lived to pass on their genes to chatty offspring. Language may be essential, but the invention of reading and writing was pure luxury. The uphill march children find in learning how to read reminds us that it may be one of our best tools, but it's not an instinct. I didn't learn to read with fluent ease until I was in college. It takes countless hours of practice to fine-tune a brain for reading. Or anything else.

Near- or farsightedness was always assumed to be hereditary. No more. In the United States, one-third of all adults are now myopic, and nearsightedness has been soaring in Europe as well. In Asia, the numbers are staggering. A recent study testing the eyesight of students in Shanghai and young men in Seoul reported that 95 percent were nearsighted. From Canberra to Ohio, one finds similar myopia, a generation of people who can't see the forest for the trees. This malady, known as "urban eyes," stems from spending too much time indoors, crouched over small screens. Our eyeballs adjust by changing shape, growing longer, which is bad news for those of us squinting to see far away. For normal eye growth, children need to play outside, maybe watching how a squirrel's nest, high atop an old hickory tree, sways in the wind, then zooming down to the runnel-rib on an individual blade of grass. Is that brown curtsey at the bottom of the yard a wild turkey or a windblown chrysanthemum?

In the past, bands of humans hunted and gathered, eyes nimble, keenly attuned to a nearby scuffle or a distant dust-mist, as they struggled to survive. Natural light, peripheral images, a long field of view, lots of vitamin D, an ever-present horizon, and a caravan of visual feedback shaped their eyes. They chipped flint and arrowheads, flayed and stitched hides, and did other close work, but not for the entire day. Close work now dominates our lives, but that's very recent, one of the Anthropocene's hallmarks, and we may evolve into a more myopic species.

Studies also show that Google is affecting our memory in chilling ways. We more easily forget anything we know we can find online, and we tend to remember where online information is located, rather than the information itself.

Long ago, the human tribe met to share food, expertise, ideas, and feelings. The keen-eyed observations they exchanged about the weather, landscape, and animals saved lives on a daily basis. Now there are so many of us that it's not convenient to sit around a campfire. Electronic campfires are the next best thing. We've reimagined space, turning the Internet into a favorite pub, a common meeting place where we can exchange knowledge or know-how or even meet a future mate. The sharing of information is fast, unfiltered, and sloppy. Our nervous systems are living in a stream of such data, influenced not just by the environment—as was the case for millennia—but abstractly, virtually. How has this changed our notion of reality? Without our brain we're not real, but when our brain is plugged into a virtual world, then that becomes real. The body remains in physical space, while the brain travels in a virtual space that is both nowhere and everywhere at once.

ONE MORNING SOME birder pals and I spend an hour at Sapsucker Woods Bird Sanctuary, watching two great blue herons feed their five rowdy chicks. It's a perfect setting for nesting herons, with an oak-snag overhanging a plush green pond, marshy shallows to hunt in, and a living larder of small fish and frogs. Only a few weeks old, the chicks are mainly fluff and appetite.

Mom and Dad run relays, and each time one returns the chicks clack wildly like wooden castanets and tussle with each other, beaks flying. Then one hogs Mom's beak by scissoring across it and holding on until a fish slides loose. The other chicks pounce, peck like speed typists, try to steal the half-swallowed fish, and if it's too late for that, grab Mom's beak and claim the next fish. Sibling rivalry is rarely so explicit. We laugh and coo like a flock of doting grandparents.

At last Mom flies off to hunt, and the chicks hush for a nap, a trial wing stretch, or a flutter of the throat pouch. Real feathers have just begun to cover their down. When a landing plane roars overhead, they tilt their beaks skyward, as if they are part of a cargo cult or expecting food from pterodactyls. We could watch their antics all day.

I'm new to this circle of blue heron aficionados, some of whom have been visiting the nest daily since April and comparing notes. "I have let a lot of things go," one says. "On purpose, though. This has been such a rare and wonderful opportunity." "Work?" another replies. "Who has time to work?"

So true. The bird sanctuary offers a rich mosaic of live and fallen trees, mallards, songbirds, red-tailed hawks, huge pileated wood-peckers, and of course yellow-bellied sapsuckers. Canada geese have been known to stop traffic (literally)—with adults serving as cross-walk guards. It's a green mansion, and always captivating.

However, we're not really there. We're all—more than 1.5 million of us thus far—watching on two live webcams affixed near the nest, and "chatting" in a swiftly scrolling Twitter-like conversation that rolls alongside the bird's-eye view.

We're virtually at the pond, without the mud, sweat, and mos-quitoes. No need to dress, share snacks, make conversation. Some of us may be taking a coffee break, or going digitally AWOL during class or work. All we can see is the heron nest up close, and that's a wonderful treat we'd miss if we were visiting on foot. In a couple of weeks the camera will follow the chicks as they learn to fish.

This is not an unusual way to pass time nowadays, and it's swiftly becoming the preferred way to view nature. Just a click away, I could have chosen a tarantula-cam, meerkat-cam, blind-mole-rat-cam, or twenty-four-hour-a-day Chinese-panda-cam from a profusion of equally appealing sites, some visited by tens of millions of people. Darting around the world to view postage-stamp-size versions of wild animals that are oblivious to the video camera is the ultimate cinema verité, and an odd shrinking and flattening of the animals, all of whom seem smaller than you. Yet I rely on virtual nature to observe animals I may never see in the wild. When I do, abraca-dabra, a computer mouse becomes a magic wand and there is an orphan wombat being fed by wildlife rescuers in Australia. Or from 308 photos of cattle posted on Google Earth I learn that herds tend to face either north or south, regardless of weather conditions, probably because they're able to perceive magnetic fields, which helps them navigate, however short the distance. Virtual nature offers views and insights that might otherwise escape us. It also helps to satisfy a longing so essential to our well-being that we feel compelled to tune in, and we find it hypnotic.

What happens when that way of engaging the world becomes habitual? Nature now comes to us, not the other way round—on a small glowing screen. You can't follow a beckoning trail, or track a noise off-camera. You don't exercise as you meander, uncertain what delight or danger may greet you, while feeling dwarfed by forces older and larger than yourself. It's a radically different way of being—*with* nature, but not *in* nature—and it's bound to shape us.

Films and TV documentaries like *Microcosmos*, *Winged Migration*, *Planet Earth*, *March of the Penguins*, and *The Private Life of Plants* inspire and fascinate millions while insinuating environmental concerns into the living room. It's mainly in such programs that we see animals in their natural settings, but they're dwarfed, flattened, interrupted by commercials, narrated over, greatly edited, and sometimes staged for added drama. Important sensory feedback is missing: the pungent mix of grass, dung, and blood; drone of flies and cicadas, dry rustling of wind through tall grass; welling of sweat; sandpapery sun.

On YouTube I just glimpsed several icebergs rolling in Antarctica —though without the grandeur of size, sounds, colors, waves, and panorama. Oddest of all, the icebergs looked a bit grainy. Lucky enough to visit Antarctica years ago, I was startled to find the air so clear that glare functioned almost as another color. I could see longer distances. Some icebergs are pastel, depending on how much air is trapped inside. And icebergs produce eerie whalelike songs when they rub together. True, in many places it's a crystal desert, but in others life abounds. An eye-sweep of busy seals, whales, penguins and other birds, plus ice floes and calving glaciers, reveals so much drama in the foreground and background that it's like entering a pop-up storybook. Watching icebergs online, or even at an Imax theater, or in sumptuous nature films, can be stirring, educational, and thought-provoking, but the experience is wildly different.

Last summer, I watched as a small screen in a department store window ran a video of surfing in California. That simple display mesmerized high-heeled, pin-striped, well-coiffed passersby who couldn't take their eyes off the undulating ocean and curling waves that dwarfed the human riders. Just as our ancient ancestors drew animals on cave walls and carved animals from wood and bone, we decorate our homes with animal prints and motifs, give our children stuffed animals to clutch, cartoon animals to watch, animal stories to read. Our lives trumpet, stomp, and purr with animal tales, such as *The Bat Poet*, *The Velveteen Rabbit*, *Aesop's Fables*, *The Wind in the Willows*, *The Runaway Bunny*, and *Charlotte's Web*. I first read these wondrous books as a grown-up, when both the adult and the kid in me were completely spellbound. We call each other by "pet" names, wear animal-print clothes. We ogle plants and animals up close on screens of one sort or another. We may not worship or hunt the animals we see, but we still regard them as necessary physical and spiritual companions. It seems the more we exile ourselves from nature, the more we crave its miracle waters. Yet technological nature can't completely satisfy that ancient yearning.

What if, through novelty and convenience, digital nature replaces biological nature? Gradually, we may grow used to shallower and shallower experiences of nature. Studies show that we'll suffer. Richard Louv writes of widespread "nature deficit disorder" among children who mainly play indoors—an oddity quite new in the history of humankind. He documents an upswell in attention disorders, obesity, depression, and lack of creativity. A San Diego fourth-grader once told him: "I like to play indoors because that's where all the electrical outlets are." Adults suffer equally. It's telling that hospital patients with a view of trees heal faster than those gazing at city buildings and parking lots. In studies conducted by Peter H. Kahn and his colleagues at the University of Washington, office workers in windowless cubicles were given flat-screen views of nature. They reaped the benefits of greater health, happiness, and efficiency than those without virtual windows. But they weren't *as* happy, healthy, or creative as people given real windows with real views of nature.

As a species, we've somehow survived large and small ice ages, genetic bottlenecks, plagues, world wars, and all manner of natural disasters, but I sometimes wonder if we'll survive our own ingenuity. At first glance, it seems like we may be living in sensory overload. The new technology, for all its boons, also bedevils us with speed demons, alluring distractors, menacing highjinks, cyber-bullies, thought-nabbers, calm-frayers, and a spiky wad of miscellaneous news. Some days it feels like we're drowning in a twittering bog of information. But, at exactly the same time, we're living in sensory poverty, learning about the world without experiencing it up close, right here, right now, in all its messy, majestic, riotous detail. Like seeing icebergs without the cold, without squinting in the Antarctic glare, without the bracing breaths of dry air, without hearing the chorus of lapping waves and shrieking gulls. We lose the salty smell of the cold sea, the burning touch of ice. If, reading this, you can taste those sensory details in your mind, is that because you've experienced them in some form before, as actual experience? If younger people never experience them, can they respond to words on the page in the same way?

The farther we distance ourselves from the spell of the present, explored by all our senses, the harder it will be to understand and protect nature's precarious balance, let alone the balance of our own human nature. I worry about our virtual blinders. Hobble all the senses except the visual, and you produce curiously deprived voyeurs. At some medical schools, future doctors can attend virtual anatomy classes, in which they can dissect a body by computer—minus that whole smelly, fleshy, disturbing *human* element. Stanford's Anatomage (formerly known as the Virtual Dissection Table) offers corpses that can be nimbly dissected from many viewpoints, plus ultrasound, X-ray and MRI. At New York University, medical students can don 3D glasses and explore virtual cadavers stereoscopically, as if swooping along Tokyo's neon-cliffed streets on Google Maps. The appeal is easy to understand. As one twenty-one-year-old female NYU student explains, "In a cadaver, if you remove an organ, you cannot add it back in as if it were never removed. Plus, this is way more fun than a textbook." Exploring virtual cadavers offers constant change, drama, progress. It's more interactive, more lively, akin to a realistic video game instead of a static corpse that just lies there.

When all is said and done, we only exist in relation to the world, and our senses evolved as scouts who work together to bridge that divide and provide volumes of information, warnings, and rewards. But they don't report everything. Or even most things. We'd collapse from sheer exhaustion. They filter experience, so that the brain isn't swamped by so many stimuli that it can't focus on what may be lifesaving. Some of our expertise comes with the genetic suit, but most of it must be learned, updated, and refined, through the fine art of focusing deeply, in the present, through the senses, and combining emotional memories with sensory experience.

Once you've held a ball, felt its smooth contour, turning it in your hands, your brain need only see another ball to remember the feel of *roundness*. You can look at a Red Delicious apple and know the taste will be sweet, the sound will be crunchy, and feel the heft of it in your hand. Strip the brain of feedback from the mansion of the senses and life not only feels poorer, learning grows less reliable. Digital exploration is predominantly visual, and nature, pixilated, is mainly visual, so it offers one-fifth of the information. Subtract the other subtle physical sensations of smell, taste, touch, and sound, and you lose a wealth of problem-solving and lifesaving detail.

When I was little, children begged to go outside and play, especially in winter when snow fell from the sky like a great big toy that clotted your mittens, whisked up your nose, slid underfoot, shape-shifted in your hands, made great projectiles, and outlined everything, linking twigs and branches, roofs and sidewalks, car hoods and snow forts with white ribbons. Some still do. But most people play more indoors now, mainly alone and stagestruck, staring at our luminous screens.

I relish technology's scope, reach, novelty, and remedies. But it's also full of alluring brain closets, in which the brain may be well occupied but has lost touch with the body, lost the intimacy of the senses, lost a visceral sense of being one life form among many on a delicately balanced planet. A big challenge for us in the Anthropocene will be reclaiming that sense of *presence*. Not to forgo high-speed digital life, but balance it with slow hours of just being outside, surrounded by nature, and watching what happens next.

Because something wonderful always happens. When a sense of presence steals up the bones, one enters a mental state where needling worries soften, careers slow their cantering, and the imaginary line between us and the rest of nature dissolves. Then for whole moments one may see nothing but snow, gathering thick and wet along the limbs of an old magnolia. Or, indoors, one may watch how a vase full of tulips, whose genes have traveled eons and silk roads, arch their spumoni-colored ruffles and nod gently when the furnace gusts. On the periodic table of the heart, somewhere between *wonderon* and *unattainium*, lies *presence*, which one doesn't so much take as steep in, like a romance, and without which one can live just fine, but not thrive. Those sensory bridges need to stay sharp, not just for our physical survival, but so we feel fully engaged and alive.

A digital identity in a digital landscape figures indelibly in our reminted sense of self. Electronic work and dreams fuel most people's lives, education, and careers. Kindness, generosity, bullying, greed, and malice all blink across our devices and survive like extremophiles on invisible nets. Sometimes, still human but mentally fused with our technologies, we no longer feel compatible with the old environment, when nature seemed truly natural. To use an antique metaphor, the plug and socket no longer fit snugly. We've grown too large, and there's no shrinking back. Instead, so that we don't feel like we're falling off the planet, we're revising and redefining nature. That includes using the Internet as we do our other favorite tools, as a way to extend our sense of self. A rake becomes an extension of one's arm. The Internet becomes an extension of one's personality and brainpower, an untethered way to move commerce and other physical objects through space, a universal diary, a stew of our species' worries, a hippocampus of our shared memories. Could it ever become conscious? It's already the sum of our daily cogitations and desires, a powerful ghost that can not only haunt with aplomb but rabble-rouse, wheel and deal, focus obsessively, pontificate on all topics, speak in all tongues, further romance, dialogue with itself, act decisively, mumble numerically, and banter between computers until the cows come home. Then find someone to milk the cows.

It's been suggested that we really have two selves now, the physical one and a second self that's always present in our absence—an online self we also have to groom and maintain, a self people can respond to even when we're not available. As a result everyone goes through two adolescences on the jagged and painfully exposed road to a sense of identity.

Surely we can inhabit both worlds with poise, dividing our time between the real and the virtual. Ideally, we won't sacrifice one for the other. We'll play outside and visit parks and wilds on foot, and also enjoy technological nature as a mental seasoning, turning to it for what it does best: illuminate all the hidden and mysterious facets of nature we can't experience or fathom on our own.

THE INTERSPECIES INTERNET

At the Toronto Zoo, Matt offers Budi one of several musical apps—a piano keyboard—and Budi stretches four long fingers through the bars and knuckle-taps an atonal chord, then several more.

"There you go! That's good!" Matt says encouragingly. "Do a couple more." One prismatic chord follows another, as Budi knuckle-dances across the iPad.

I'm reminded of the YouTube video in which Panbanisha, a nineteen-year-old bonobo at the Language Research Center in Atlanta, is introduced to a full-size keyboard for the first time by the musician Peter Gabriel. Sitting on the piano bench, she considers the keyboard for a moment, then noodles around on it, discovers a note she likes, then finds the octave and picks out notes within it, creating a melody that floats above Gabriel's improvised background. Especially wondrous is her sense of musical timing, the negative space between notes when, neither rushed nor dragged, each note hovers in the air like a diver at the arc of a dive, before falling into a shared pool of reverberating silence, from which, at a pleasing interval, another note arises. After a while, she cuts loose and jams harmonies with his vocals.

"There was clear, sharp, musical intelligence at work," Gabriel says. She was "tender and open and expressive."

Her brother Kanzi came in next, and even though he'd never sat at a piano before, when he saw how much attention his sister was getting, "he threw down his blanket like James Brown discarding one of his cloaks," Gabriel says, "and then does this, you know, fantastic sort of triplet improvisation."

Gabriel finds orangutans the bluesmen of the ape world, "who always look a little sad but they're amazingly soulful."

At seven, Budi is still a kid, not a bluesman, and he enjoys playing memory and cognitive games on the iPad, or using the musical and drawing apps, but he's most fascinated by YouTube videos of other orangutans.

Matt explains, tenderly, that he believes in offering orangutans a way to communicate nonverbally with other apes, including us. Keepers could always hand them things, but if the orangs "could tell anybody what they want, then their lives would get a lot more fulfilling."

The most ambitious version of that desire is known as the Interspecies Internet. Matt has heard of it, and thinks it would be a cool thing to do, though the logistics might be tough. Ever since the 1980s, the cognitive psychologist Diana Reiss, who studies animal intelligence, has been teaching dolphins to use an underwater keyboard (soon to be replaced with a touchscreen) to ask for food, toys, or favorite activities. She and the World Wide Web pioneer (and Chief Internet Evangelist at Google) Vent Cerf, Peter Gabriel, and Neil Gershenfeld, director of MIT's Center for Bits and Atoms, are combining their wide-ranging talents to launch a touchscreen network for cockatoos, dolphins, octopuses, great apes, parrots, elephants, and other intelligent animals to communicate directly with humans and each other.

When the four introduced the idea to the world at a TED Talk,

Gabriel said: "Perhaps the most amazing tool man has created is the Internet. What would happen if we could somehow find new interfaces—visual, audio—to allow us to communicate with the remarkable beings we share the planet with?" He told of his great respect for the intelligence of apes, and how, growing up on a farm in England, he used to peer into the eyes of cattle and sheep and wonder what they were thinking.

In response to those who say, "The Internet is dehumanizing us. Why are we imposing it on animals?" Gabriel replied: "If you look at a lot of technology, you'll find that the first wave dehumanizes. The second wave, if it's got good feedback and smart designers, can superhumanize." He'd love for any intelligent species that is interested to explore the Internet in the same way we do.

Cerf added that we shouldn't restrict the Internet to one species. Other sentient species should be part of the network, too. And, in that spirit, the most important aspect of the project is learning how to communicate with species "who are not us but share a sensory environment."

Gershenfeld said that when he saw the video clip of Panbanisha jamming with Gabriel, he was struck by the history of the Internet. "It started as the Internet of mostly middle-aged white men," he said. "I realized that we humans had missed something—the rest of the planet."

If the Interspecies Internet is the next logical step, what will it be a prelude to? Gershenfeld looks forward to "computers without keypads or mice," controlled by reins of thought, prompted by waves of feelings and memories. It's one thing to be able to translate our ideas into the physical environment, but a giant step for humankind to do that with thoughts alone. Telekinesis used to belong only to science fiction, but we're well on our way to that ascendancy now, as paralyzed patients learn to wield prosthetic arms and propel exoskeleton legs via muscular thoughts. These possibilities change how we imagine the brain, no longer a skull-bound captive.

"Forty years ago," Cerf said, "we wrote the script of the Internet.

Thirty years ago we turned it on. We thought we were building a system to connect computers together. But we quickly learned that it's a system for connecting people." Now we're "figuring out how to communicate with something that's not a person. You know where this is going," Cerf continued. "These actions with other animals will teach us, ultimately, how we might interact with an alien species from another world. I can hardly wait." Cerf is leading a NASA initiative to create an Interplanetary Internet, which can be used by crews on spacecraft between the planets. Who knows what spin-off Internets will follow.

Reiss pointed out that dolphins are mighty alien. "These are true *non*terrestrials."

The Apps for Apes program is but one part of our postindustrial, nanotech, handcrafted, digitally stitched world in which luminous webs help us relate to friends, strangers, and other intelligent life forms, whether or not they have a brain.

YOUR PASSION FLOWER
IS SEXTING YOU

L ife takes many forms, as does *intelligence*—plants may not possess a brain, but they can be diabolically clever, manipulative, and vicious. So it was only a matter of time. Plants have begun texting for help. Thanks to clever new digital devices, a dry philodendron, undernourished hibiscus, or sadly neglected wandering Jew can either text or tweet to its owner over the Internet. Humans like to feel appreciated, so a begonia may also send a simple "Thank you" text—when it's *happy*, as gardeners like to say, meaning healthy and well tended. Picture your Boston fern home alone placing botani*calls*. But why should potted plants be the only ones to reassure their humans? Another company has found a way for crops to send a text message in unison, letting their farmer know if she's doing a good enough job to deserve a robust harvest. Sensors lodged in the soil respond to moisture and send prerecorded messages customized by the owner. What is the sound of one hand of bananas clapping?

Plants texting humans may be new, but malcontent plants have always been chatting among themselves. When an elm tree is being attacked by insects, it does the chemical equivalent of broadcasting *I'm hurt! You could be next!* alerting others in its grove to whip up some dandy poisons. World-class chemists, plants vie with Lucrezia Borgia dressed in green. If a human kills with poison, we label it a wicked and premeditated crime, one no plea of "self-defense" can excuse. But plants dish out their nastiest potions every day, and we wholeheartedly forgive them. They may lack a mind, or even a brain, but they do react to injury, fight to survive, act purposefully, enslave humans (through the likes of coffee, tobacco, opium), and gab endlessly among themselves.

Strawberry, bracken, clover, reeds, bamboo, ground elder, and lots more all grow their own social networks—delicate runners (really horizontal stems) linking a grove of individuals. If a caterpillar chews on a white clover leaf, the message races through the colony, which ramps up its chemical weaponry. Stress a walnut tree and it will brew its own caustic aspirin and warn its relatives to do the same. Remember Molly Ivins's needle-witted quip about an old Texan congressman: "If his IQ slips any lower, we'll have to water him twice a day"? She clearly misjudged the acumen of plants. Plants are not mild-mannered. Some can be murderous, seductive, deceitful, venomous, unscrupulous, sophisticated, and downright barbaric.

Since they can't run after a mate, they go to phenomenal lengths to con animals into performing sex for them, using a vaudeville trunk full of costumes. For instance, some orchids disguise themselves as the sex organs of female bees so that male bees will try to mate with them and leave wearing pollen pantaloons. Since they can't run from danger, they devise a pharmacopeia of poisons and an arsenal of simple weapons: hideous killers like strychnine and atropine; ghoulish blisterers like poison ivy and poison sumac; slashers like holly and thistle waving scalpel-sharp spines. Blackberries and roses wield belts of curved thorns. Each hair of a stinging nettle brandishes a tiny syringe full of formic acid and histamine to make us itch or run.

Just in case you're tempted to cuddle your passion flower when you teach it to send text messages—resist the urge. Passion flowers release cyanide if their cell walls are broken by a biting insect or a fumbling human. Of course, because nature is often an arms race, leaf-eating caterpillars have evolved an immunity to cyanide. Not us, alas. People have died from accidentally ingesting passion flower, daffodils, yew, autumn crocuses, monkshood, rhododendron, hyacinths, peace lilies, foxglove, oleander, English ivy, and the like. And one controversial theory about the Salem witch trials is that the whole shameful drama owes its origin to an especially wet winter when the rye crop was infected with ergot, an LSD-like hallucinogen that, perhaps breathed in by those grinding it into flour, caused women to act bewitched.

Today we're of two minds about undisciplined plants just as we are about wild animals. We want them everywhere around us, but not roaming freely. We keep pet plants indoors or outside, provided they're well behaved and don't run riot. Weeds alarm us. And yet, as Patrick Blanc points out, "it is precisely this form of freedom of the plant world that most fascinates us." Devious and dangerous as plants can be, they adorn every facet of our lives, from courtship to burial. They fill our rooms with piquant scents, dazzling tableaux, and gravity-defying aerial ballets and contortions as they unfold petals and climb toward the sun. Think of them as the original Cirque du Soleil. Many an African violet has given a human shrinking violet a much-needed interkingdom friendship.

Since they do demand looking after, and we do love our social networks, I expect texting will sweep the plant world, showering us with polite thank-yous and rude complaints. What's next, a wisteria sexting every time it's probed by a hummingbird? A bed of zinnias ranting to online followers as they go to seed?

Surely some playful wordsmiths need to dream up spirited texts for the botanicalling plants to send, telegrams of fulsome fawning or sarcastic taunt. Maybe a little soft soap: "You grow girl! Thanks for the TLC." Or think how potent it would be, in the middle of a dinner date, to receive a text from your disgruntled poinsettia that reads: "With fronds like you who needs anemones?!"

WHEN ROBOTS WEEP,
WHO WILL COMFORT THEM?

t's an Anthropocene magic trick, this extension of our digital selves over the Internet, far enough to reach other people, animals, plants, interplanetary crews, extraterrestrial visitors, the planet's Google-mapped landscapes, and our habitats and possessions. If we can revive extinct life forms, create analog worlds, and weave new webs of communication—what about new webs of life? Why not synthetic life forms that can sense, feel, remember, and go through Darwinian evolution?

HOD LIPSON IS the only man I know whose first name means "splendor" in Hebrew and a V-shaped wooden trough for carrying bricks over one shoulder in English. The paradox suits him physically and mentally. He looks strong and solid enough to carry a hod full of bricks, but he would be the first to suggest that the bricks might not resemble any you've ever known. They might even saunter, reinvent themselves, refuse to be stacked, devise their own mortar, fight back, explore, breed more of their kind, and boast a nimble curiosity about the world. Splendor can be bricklike, if graced by complexity.

His lab building at Cornell University is home to many a skunk-works project in computer sciences or engineering, including some of DARPA's famous design competitions (agile robots to clean up toxic disasters, superhero exoskeletons for soldiers, etc.). Nearby, two futuristic DARPA Challenge cars have been left like play-worn toys a few steps from a display case of antique engineering marvels and an elevator that's old and slow as a butter churn.

On the second floor, a black spider-monkey-like robot clings to the top left corner of Lipson's office door, intriguing but inscrutable, except to the inner circle for whom it's a wry symbol and tradesman's sign of the sort colonial shopkeepers used to hang out to identify their business: the apothecary's mortar and pestle, the chandler's candles, the cabinetmaker's hickory-spindled armchair, the roboticist's apprentice. Though in its prime the leggy bot drew the keen gaze of students, students come and go, as do the smart-bots they work on, which, coincidentally, seem to have a life span of about 3.5 years—how long it takes a student to finish a dissertation and graduate.

A man with curly hair, chestnut-brown eyes, and a dimpled chin, Hod welcomes me into his cheerful office: tall windows, a work desk, a Dell computer with a triptych of screens, window boxes for homegrown tomatoes in summer, and a wall of bookshelves, atop which sits an array of student design projects. To me they look unfamiliar but strangely beautiful and compelling, like the merchandise in an extraterrestrial bazaar. A surprisingly tall white table and its chairs invite one to climb aboard and romp with ideas. At Lipson's round table, if you're under six feet tall, your feet will automatically leave the planet, which is good, I think, because even this limited levitation aids the imagination, untying gravity just enough to make magic carpet rides, wing-walkers, and spaceships humble as old rope. There's a reason we cling to such elevating turns of phrase as "I was walking on air," "That was uplifting," "heightened awareness," "surmounting obstacles," or "My feet never touched the ground." The mental mischief of creativity—which thrives on such fare as deep play, risk, a superfluity of ideas, the useful application of obsession, and willingly backtracking or hitting dead ends without losing heart—is also fueled by subtle changes in perception. So why not cast off mental moorings and hover a while each day?

What's the next hack for a rambunctious species full of whiz kids with digital dreams? Lipson is fascinated by a different branch of the robotic evolutionary tree than the tireless servant, army of skilled hands, or savant of finicky incisions with which we have become familiar. Over ten million Roomba vacuum cleaners have already sold to homeowners (who sometimes find them being ridden as child or cat chariots). We watch with fascination as robotic sea scouts explore the deep abysses (or sunken ships), and NOAA's robots glide underwater to monitor the strength of hurricanes. Google's robotics division owns a medley of firms, including some minting life-size humanoids—because, in public spaces, we're more likely to ask a cherub-faced robot for info than a touchscreen. Both Apple and Amazon are diving into advanced robotics as well. The military has invested heavily in robots as spies, bionic gear, drones, pack animals, and bomb disposers. Robots already work for us with dedicated precision in factory assembly lines and operating rooms. In cross-cultural studies, the elderly will happily adopt robotic pets and even babies, though they aren't keen on robot caregivers at the moment.

All of that, to Lipson, is child's play. His focus is on a self-aware species, *Robot sapiens*. Our own lineage branched off many times from our apelike ancestors, and so will the flowering, subdividing lineage of robots, which perhaps needs its own Linnaean classification system. The first branch in robot evolution could split between AI and AL—artificial intelligence and artificial life. Lipson stands right at that fork in that road, whose path he's famous for helping to divine and explore in one of the great digital adventures of our age. It's the ultimate challenge, in terms of engineering, in terms of creation.

"At the end of the day," he says with a nearly illegible smile, "I'm trying to recreate *life* in a synthetic environment—not necessarily something that will look human. I'm not trying to create a person who will walk out the door and say 'Hello!' with all sorts of anthropomorphic features, but rather features that are truly alive given the principles of life—traits and behaviors they have evolved on their own. I don't want to build something, turn it on, and suddenly it will be alive. I don't want to *program* it."

A lot of robotics today, and a lot of science fiction, is about a human who schemes at a workbench in a dingy basement, digitally darning scraps, and then figuring out how to command his scarecrow to do his bidding. Or a mastermind who builds the perfect robots that eventually go haywire in barely discernible stages and start to massacre us, sometimes on Earth, often in space. It assumes an infinite power that humans have (and so can lose) over the machine.

Engineering's orphans, Lipson's brainchildren would be the first generation of truly self-reliant machines, gifted with free will by their soft, easily damaged creators. These synthetic souls would fend for themselves, learn, and grow—mentally, socially, physically—in a body not designed by us or by nature, but by fellow computers.

That may sound sci-fi, but Lipson is someone who relishes not only pushing the envelope but tinkering with its dimensions, fabric, inertia, and character. For instance, bothered by a question that nags sci-fi buffs, engineers, and harried parents alike—*Where are all the robots we were told would be working for us by now?*—he decided to go about robotics in a new way. And also in the most ancient of ways, by summoning the "mother of all designers, Evolution," and asking a primordial soup of robotic bits and pieces to zing through millions of generations of fluky mutations, goaded by natural selection. Of course, natural evolution is a slapdash and glacially slow mother, yielding countless bottlenecks for every success story. But computers can be programmed to "evolve" at great speed with digital finesse, and adapt to all the rigors of their environment.

Would they be able to taste and smell? I wonder, realizing at once how outmoded the very question is. Taste buds rise like flaky volcanoes on different regions of the tongue, with bitter at the back, lest we swallow poisons. How hard would it be to evolve a suite of specialized "taste buds" that bear no resemblance to flesh? Flavor engineers at Nestlé in Switzerland have already created an electronic "taster" of espresso, which analyzes the gas different pulls of ristretto give off when heated, translating each bouquet of ions into such human-friendly, visceral descriptions as "roasted," "flowery," "woody," "toffee," and "acidy."

However innovative, Lipson's entities are still primitive when compared to a college sophomore or a bombardier beetle. But they're the essential groundwork for a culture, maybe a hundred years from now, in which some robots will do our bidding, and others will share our world as a parallel species, one that's creative and curious, moody and humorous, quick-witted, multitalented, and 100 percent synthetic. Will we regard them as *life*, as a part of nature, if they're not carbon-based—as are all of Earth's plants and animals? Can they be hot-blooded without blood? How about worried, petulant, sly, envious, downright cussed? The future promises fleets of sovereign silicants and, ultimately, self-governing, self-reliant robotic angels and varmints, sages and stooges. To be able to ponder such possibilities is a testament to the infinite agility of matter and its great untapped potential.

Whenever Lipson talks of robots being truly *alive*, gently stressing the word, I don't hear Dr. Frankenstein speaking, at one in the morning, as the rain patters dismally against the panes,

> when, by the glimmer of the half-extinguished light, I saw the dull yellow eye of the creature open; it breathed hard, and a convulsive motion agitated its limbs. How can I describe my emotions at this catastrophe, or how delineate the wretch whom with such infinite pains and care I had endeavoured to form?

As in the book's epigraph, lines from Milton's *Paradise Lost*: "Did I request thee, Maker, from my clay / To mould Me man?" Mary Shelley suggests that the parent of a monster is ultimately responsible for all the suffering and evil he has unleashed. From her early years of seventeen to twenty-one, Shelley was herself consumed by physical creation and literally sparking life, becoming pregnant and giving birth repeatedly, only to have three of her four children die soon after birth. She was continually pregnant, nursing, or mourning—creating and being depleted by her own creations. That complex visceral state fed her delicately horrifying tale.

In her day, scientists were doing experiments in which they animated corpses with electricity, fleetingly bringing them back to life, or so it seemed. Whatever the image of Frankenstein's monster may have meant to Shelley, it has seized the imagination of people ever since, symbolizing something unnatural, Promethean, monstrous that we've created by playing God, or from evil motives or through simple neglect (Dr. Frankenstein's sin wasn't in creating the monster but in abandoning it). Something we've created that, in the end, will extinguish us. And that's certainly been a key theme in science-fiction novels and films about robots, androids, golems, zombies, and homicidal puppets. Such ethical implications aren't Lipson's concern; that's mainly for seminars and summits in a future he won't inhabit. But such discussions are already beginning on some campuses. We've entered the age of such college disciplines as "robo-ethics" and Lipson's specialty, "evolutionary robotics."

Has it come to this, I wonder, creating novel life forms to prove we can, because a restless mind, left to its own devices and given enough time, is bound to create equally restless devices, just to see what happens? It's a new threshold of creators creating creative beings.

"Creating life is certainly a tall pinnacle to surmount. Is it also a bit like having children?" I ask Lipson.

"In a different way. . . . Having children isn't so much an intellectual challenge, but other kinds of challenges." His eyebrows lift slightly to underline the understatement, and a memory seems to flit across his eyes.

"Yes, but you set them in motion and they don't remake themselves exactly, but . . ."

"You have very little control. You can't program a child . . ."

"But you can shape its brain, change the wiring."

"Maybe you can shape some of the child's experiences, but there are others you can't control, and a lot of the personality is in the genes: nature, not nurture. Certainly in the next couple of decades we won't be programming machines, but . . . like children, exactly . . . we'll shape their experiences a little bit, and they'll grow on their own and do what they do."

"And they'll simply adjust to whatever job is required?"

"Exactly. Adaptation and adjustment, and with that will come other issues, and a lot of problems." He smiles the smile of someone who has seen dust-ups on a playground. "Emotions will be a big part of that."

"You think we'll get to the point where machines have deep emotions?"

"They will have deep emotions," Hod says, certain as the tides. "But they won't necessarily be *human* emotions. And also machines will not always do what we want them to do. This is already happening. Programming something is the ultimate control. You get to make it do exactly what *you* want *when* you want it. This is how robots in factories are programmed to work today. But the more we give away some of our control over how the machine learns . . ."

As a cool gust of October air wafts through the screenless window, carrying a faint scent of crumbling magnolia leaves and damp earth, it trails gooseflesh across my wrist.

"Let me close the window." Hod slides gingerly off the tall chair as if from a soda fountain seat and closes the gaping mouth of the window.

We were making eye contact; how did he notice my gooseflesh? Stare at something and only the center of your vision is in focus; the periphery blurs. Is his visual compass wider than most people's, or is he just being a thoughtful host and, sensing a breeze himself, reasoning that since I'm sitting closer to the window I might be feeling chillier? As we talk, his astonishingly engineered biological brain—with its flexible, self-repairing, self-assembling, regenerating components that won't leave toxic metals when they decompose—is working hard on several fronts: picturing what he wants to say in all of its complexity; rummaging through a sea of raw and thought-rinsed ideas; gauging my level of knowledge—very low in his field; choosing the best way to translate his thoughts into words for this newly met and unfamiliar listener; reading my unconscious cues; rethinking some of his words when they're barely uttered; revising them right as they're leaving his mouth, in barely perceptible changes to a word's opening sound; choosing the ones most accurate on several levels (literally, professionally, emotionally, intellectually) whose meaning I may nonetheless give subtle signs of not really understanding—signs visible to him though unconscious to me, as they surface from a dim warehouse of my previous thoughts and experiences and a vocabulary in which each word carries its own unique emotional valence—while at the same time he's also forming impressions of me, and gauging the impression I might be forming of him . . .

This is called a "conversation," the spoken exchange of thoughts, opinions, and feelings. It's hard to imagine robots doing the same on as many planes of meaning, layered emotions, and spring-loaded memories.

Beyond the windows with their magenta-colored accordion blinds, and the narrow Zen roof garden of rounded stones, twenty yards across the courtyard and street, behind a flimsy orange plastic fence, giant earth-diggers and men in hard hats are tearing up rock and soil with the help of machines wielding fierce toothy jaws. Such brutish dinosaurs will one day give way to rational machines that can transform themselves into whatever the specific task requires— perhaps the sudden repair of an unknown water pipe—without a boss telling them what to do. By then the din of jackhammers will also be antiquated, though I'm sure our hackles will still twitch at the scrape of clawlike metal talons on rock.

"When a machine learns from experience, there are few guarantees about whether or not it will learn what you want," Lipson continues as he remounts his chair. "And it might learn something that you didn't want it to learn, and yet it can't forget. This is just the beginning."

I shudder at the thought of traumatized robots.

He continues, "It's the unspoken Holy Grail of a lot of roboticists— to create just this kind of self-awareness, to create consciousness."

What do roboticists like Lipson mean when they speak of "conscious" robots? Neuroscientists and philosophers are still squabbling over how to define consciousness in humans and animals. On July 7, 2012, a group of neuroscientists met at the University of Cambridge to declare officially that nonhuman animals "including all mammals and birds, and many other creatures, including octopuses" are conscious. To formalize their position, they signed a document entitled "The Cambridge Declaration on Consciousness in Non-Human Animals."

But beyond being conscious, humans are quintessentially self-aware. Some other animals—orangutans and other cousins of ours, dolphins and octopuses, and some birds—are also self-aware. A wily jay might choose to cache a seed more quietly because other jays are nearby and it doesn't want the treasure stolen; an octopus might take the lid off its habitat at night to go for a stroll and then replace the lid when it returns lest its keepers find out. They possess a theory of mind, and can intuit what a rival might do in a given situation and act accordingly. They exhibit deceit, compassion, the ability to see themselves through another's eyes. Chimpanzees feel deeply, strategize, plan, think abstractly to a surprising degree, mourn, empathize some, deceive, seduce, and are all too conscious of life's pressures, if not its chastening illusions. They're blessed and burdened, as we are, by strong family ties and quirky personalities, from madcap to martinet. They jubilate when happy, mope when sad.

I don't think they fret and reason endlessly about mental states, as we do. They simply dream a different dream, probably much like the one we used to dream, before we crocheted into our neural circuitry the ability to have ideas about everything. Other animals may know you know something, but they don't know you know they know. Other mammals may think, but we think about having thoughts. Linnaeus categorized us in the subspecies of *Homo sapiens sapiens*, adding the extra *sapiens* because we don't just know, we *know* that we know. Our infants respond to their surroundings and other people, and start evolving a sense of self during their first year. Like orangutans, elephants, and even European magpies, they can identify themselves in a mirror, and they gather that others have a personal point of view that differs from their own.

So when people talk about robots being conscious and self-aware, they mean a range of knowing. Some robots may be smarter than humans, more rational, more skillful in designing objects, and better at anything that requires memory and computational skills. I reckon they can be deeply curious (though not exactly the way we are), and will grow even more so. They can already do an equivalent of what we think of as ruminating and obsessing, though in fewer dimensions. Engineers are designing robots with the ability to attach basic feelings to sensory experience, just as we do, by interacting with the world, filing the memory, and using it later to predict the safety of a situation or the actions of others.

Lipson wants his robots to make assumptions and deductions based on past experiences, a skill underlying our much-prized autobiographical memory, and an essential component of learning. Robots will learn through experience not to burn a hand on a hot stove, and to look both ways when crossing the street. There are also subtle, interpersonal clues to decipher. For instance, Lipson uses the British "learnt" instead of the American "learned," but the American "while" instead of the British "whilst." So, from past experience, I deduce that he learned English as a child from a British speaker, and assume he has lived in the United States just long enough to rinse away most of the British traces.

Yet however many senses robots may come to possess—and there's no reason why they shouldn't have many more than we, including sharper eyesight and the ability to see in the dark—they'll never be embodied exactly like us, with a thick imperfect sediment of memories, and maybe a handful of diaphanous dreams. Who can say what unconscious obbligato prompts a composer to choose this rhythm or that—an irregular pounding heart, tinnitus in the ears, a lover who speaks a foreign language, fond memories evoked by the crackle of ice in winter, or an all too human twist of fate? There would be no *Speak, Memory* from Nabokov, or *The Gulag Archipelago* from Solzhenitsyn, without the sentimental longings of exile. I don't know if robots will be able to do the sort of elaborate thought experiments that led Einstein to discoveries and Dostoevsky to fiction.

Yet robots may well create art, from who knows what motive, and enjoy it based on their own brand of aesthetics, satire (if they enjoy satire), or humor. We might enjoy it, too, especially if it's evocative of work by human artists, if it appeals to our senses. Would we judge it differently? For one of its gallery shows, Yale's art museum accepted paintings inspired by Robert Motherwell, only to change its mind when it learned they'd been painted by a robot in Lipson's Creative Machines Lab. It would be fun to discover robots' talents and sensibility. Futurologists like Ray Kurzweil believe, as Lipson does, that a race of conscious robots, far smarter than we, will inhabit Earth's near-future days, taking over everything from industry, education, and transportation to engineering, medicine, and sales. They already have a foot in the door.

At the 2013 Living Machines Conference, in London, the European RobotCub Consortium introduced their iCub, a robot that has naturally evolved a theory of mind, an important milestone that develops in children at around the age of three or four. Standing

about three feet tall, with a bulbous head and pearly white face, programmed to walk and crawl like a child, it engages the world with
humanlike limbs and joints, sensitive fingertips, stereo vision, sharp
ears, and an autobiographical memory that's split like ours into the
episodic memory of, say, skating on a frozen pond as a child and
the semantic memory of how to tilt the skate blades on edge for a
skidding stop. Through countless interactions between body and
world it codifies knowledge about both. None of that is new. Nor
is being able to distinguish between self and other, and intuit the
other's mental state. Engineers like Lipson have programmed that
discernment into robots before. But this was the first time a robot
evolved the ability *all by itself.* iCub is just teething on consciousness,
to be sure, but it's intriguing that the bedrock of empathy, deception,
and other traits that we regard as *conscious* can accidentally emerge
during a robot's self-propelled Darwinian evolution.

It happened like this. iCub was created with a double sense of self.
If he wanted to lift a cup, his first self told his arm what to do, while
predicting the outcome and adjusting his knowledge based on whatever happened. His second—we can call it "interior"—self received
exactly the same feedback, but, instead of acting on the instructions,
it could only try to predict what would happen in the future. If the
real outcome differed from a prediction, the interior self updated its
cavernous memory. That gave iCub two versions of itself, an active
one and an interior "mental" one. When the researchers exposed
iCub's mental self to another robot's actions, iCub began intuiting
what the other robot might do, based on personal experience. It saw
the world through another's eyes.

As for our much-prized feats of scientific reasoning and insight,
Lipson's lab has created a Eureqa machine, a computer scientist
able to make a hypothesis, design an experiment, contemplate the
results, and derive laws of nature from them. Plumbing the bottom-
less depths of chaos, it divines meaning. Assigned a problem in New-
tonian physics (how a double pendulum works), "the machine took
only a couple of hours to come up with the basic laws of motion,"

Lipson says, "a task that occupied Newton for years after he was inspired by an apple falling from a tree."

Eureqa takes its name from a legendary moment in the annals of science, two thousand years ago, when Archimedes—already a renowned mathematician and inventor with formidable mastery in his field—was soaking in his bathtub, his senses temporarily numbed by warm water and weightlessness, and the solution to a problem came to him in a flash of insight. Leaping from the tub, he supposedly ran naked through the streets of Syracuse yelling, "Eureka!" ("I have found it!")

For two thousand years, that's how traditional science has run: solid learning and mastery, then the kindling of observation and a spark of insight. The Eureqa machine marks a turning point in the future of how science is done. Once upon a time, Galileo studied the movement of the heavenly bodies, Newton watched an apple fall in his garden. Today science is no longer that simple because we wade through oceans of information, generate vast amounts of additional data, and analyze it on an unprecedented scale. Virtuoso number-crunchers, our computers can extract data without bias, boredom, vanity, selfishness, or greed, quickly doing the work that used to take one human a lifetime.

In 1972, when I was writing my first book, *The Planets: A Cosmic Pastoral*, a suite of scientifically accurate poems based on the planets, I used to hang out in the Space Sciences Building at Cornell. The astronomer Carl Sagan was on my doctoral committee, and he kindly gave me access to NASA photographs and reports. At that time, it was possible in months to learn nearly everything humans knew about the other planets, and the best NASA photos of the outermost planets were only arrows pointing to balls of light. Over the decades, I attended flybys at the Jet Propulsion Laboratory in Pasadena, California, and watched the first exhilarating images roll in from distant worlds as *Viking* and *Voyager* reached Mars, Jupiter, Saturn, Neptune, and an entourage of moons. In the 1980s, it was still possible for an amateur to learn everything humans knew about the planets. Today that's no longer so. The Alps of raw data would take more than one lifetime to summit, passing countless PhD dissertations at campsites along the trail.

But all that changes with a tribe of Eureqa-like machines. A team of scientists at the University of Aberystwyth, led by Professor Ross King, has revealed the first machine able to deduce new scientific knowledge about nature on its own. Named Adam, the two-armed robot designed and performed experiments to investigate the genetics of baker's yeast. Carrying out every stage of the scientific process by itself without human intervention, it can perform a thousand experiments a day and make discoveries.

More efficient science will solve modern society's problems faster, King believes, and automation is the key. He points out that "automation was the driving force behind much of the nineteenth- and twentieth-century progress." In that spirit, King's second-generation laboratory robot, named Eve, is even faster and nimbler than Adam. It's easy to become mesmerized watching a webcam of Eve testing drugs, her automated arms and stout squarish body shuffling trays, potions, and tubes with tireless precision, as she peers through ageless nonblinking eyes, while saving the sanity of countless graduate students, spared sleepless nights in the lab tending repetitive experiments.

How extraordinary that we've created peripheral brains to discover the truths about nature that we seek. We're teaching them how to work together calmly as a society, share data at lightning speed, and cooperate so much better than we do, rubbing brains together in the invisible drawing room we sometimes call the "cloud." Undaunted, despite our physical and mental limitations, we design robots to continue the quest we began long ago: making sense of nature. Some call it Science, but it's so much larger than one discipline, method, or perspective.

I find it touchingly poetic to think that as our technology grows more advanced, we may grow more human. When labor, science, manufacturing, sales, transportation, and powerful new technologies are mainly handled by savvy machines, humans really won't be able to compete in those sectors of the economy. Instead we may dominate an economy of interpersonal or imaginative services, in which our human skills shine.

Smart robots are being nurtured and carefully schooled in laboratories all over the world. Thus far, Lipson's lab has programmed machines to learn things unassisted, teaching themselves the basic skills of how to walk, eat, metabolize, repair wounds, grow, and design others of their kind. At the moment, no one robot can do everything; each pursues its own special destiny. But one day, *all* the lab machines will merge into a single stouthearted . . . *being*—what else would we call it?

One of Lipson's robots knows the difference between self and other, the shape of its physique, and whether it can fit into odd spaces. If it loses a limb, it revises its self-image. It senses, recollects, keeps updating its data, just as we do, so that it can predict future scenarios. That's a simple form of self-awareness. He's also created a machine that can picture itself in various situations—very basic thought experiments—and plan what to do next. It's starting to think about thinking.

"Can I meet it?" I ask.

His eyes say: *If only*.

Leading me across the hall, into his lab, he stops in front of a humdrum-looking computer on a desk, one of many scattered around the lab.

"All I can show you is this ordinary-looking computer," he says. "I know it doesn't look exciting because the drama is unfolding in the software inside the machine. There's another robot," he says, gesturing to a laptop, "that can look at a second robot and try to infer what that other robot is thinking, what the other robot is going to do, what the other robot might do in a new situation, based on what it did in a previous situation. It's learning another's *personality*. These are very simple steps, but they're essential tools as we develop this technology. And with this will come emotions, because emotions, at the end of the day, have to do with the ability to project yourself into different situations—fear, various needs—and anticipate the rewards and pain in many future dramas. I hope that, as the machines learn, eventually they'll produce the same level of emotions as in humans. They might not be the same *type* of emotions, but they will be as complex and rich as in humans. But it will be different, it will be alien."

I'm fascinated by the notion of "other types of emotions." What would a synthetic species be like without all the lavish commotion of sexual ardor, wooing, jealousy, longing, affectionate bonds, shared experiences? Just as I long to know about the inner (and outer) lives of life forms on distant planets, I long to know about the obsessions, introspections, and emotional muscles that future species of robots might wrestle with. A powerful source of existential grief comes from accepting that I won't live long enough to find out.

"Emotional robots . . . I've got a hunch this isn't going to happen in my lifetime." I'm a bit crestfallen.

"Well, it will probably take a century, but that's a blip in human history, right?" he says in a reassuring tone. "What's a century? It's nothing. If you look at the curve of humans on Earth," he says, curving one hand a few inches off the table, "we're right there. That's a hundred years."

"So much has happened in just the last two hundred years," I say, shaking my head. "It's been quite an express ride."

"Exactly. And the field is accelerating. But there's good and bad, right? If you say 'emotions,' then you have depression, you have deception, you have creativity and curiosity—creativity and curiosity we're already seeing here in various machines.

"My lab is called the Creative Machines Lab because I want to make machines which *are* creative, and that's a very very controversial topic in engineering, because most engineers—*close the door, speak quietly*—are stuck in the Intelligent Design way of thinking, where the engineer is the intelligent person and the machines are *being created*, they just do the menial stuff. There's a very clear division. The idea that a *machine* can create things—possibly more creatively than the engineer that designed the machine—well, it's very troubling to some people, it questions a lot of fundamentals."

Will they grow attached to others, play games, feel empathy, crave mental rest, evolve an aesthetics, value fairness, seek diversion, have fickle palates and restless minds? We humans are so far beyond the Greek myth of Icarus, and its warning about overambition (father-and-son inventors and wax wings suddenly melting in the sun). We're now strangers in a strange world of our own devising, where becoming a creator, even the Creator, of other species is the ultimate intellectual challenge. Will our future robots also design new species, bionts whose form and mental outlook we can't yet imagine?

"What's this?" I ask, momentarily distracted by a wad of plastic nestled on a shelf.

He hands me the strange entanglement of limbs and joints, a small robot with eight stiff black legs that end in white ball feet. The body is filamental, like a child's game of cat's cradle gone terribly wrong, and it has no head or tentacles, no bulging eyes, no seedlike brain. It wasn't designed as an insect. Or designed by humans, for that matter.

Way back in our own evolution, we came from fish that left the ocean and flopped from one puddle to another. In time they evolved legs, a much better way to get around on land. When Lipson's team asked a computer to invent something that could get from point A to point B—without programming it how to walk—at first it created robots reminiscent of that fish, with multihinged legs, flopping forward awkwardly. A video, posted on YouTube, records its first steps, with Lipson looking on like a proud parent, one who appreciates how remarkable such untutored trials really are. Bits of plastic were urged to find a way to combine, think as one, and move themselves, and they did.

In another video, a critter trembles and skitters, rocks and slides. But gradually it learns to coordinate its legs and steady its torso, inching forward like a worm, and then walking insectlike—except that it wasn't told to model an insect. It dreamt up that design by itself, as a more fluent way forward. Awkward, but effective. Baby steps were fine. Lipson didn't expect grace. He could make a spider robot that would run faster, look better, and be more reliable, but that's not the point. Other robots are bending, flexing, and running, using replica tendons and muscles. DARPA's "cheetah" was recently clocked at a tireless 30 mph sprint. But that cheetah was programmed; it would be a four-legged junkpile without a human telling it what to do. Lipson wants the robot to do everything on its own, eclipsing what any human can design, unfettered by the paltry ideas of its programmers.

It's a touching goal. Surpassing human limits is so human a quest, maybe the most ancient one of all, from an age when dreams were omens dipped in moonlight, and godlike voices raged inside one's head. A time of potent magic in the landscape. Mountains attracted rain clouds and hid sacred herbs, malevolent spirits spat earthquakes or drought, tyrants ruled certain trees or brooks, offended waterholes could ankle off in the night, and most animals parleyed with at least one god or demon. What was human agency compared to that?

ROBOTS ON A DATE

Looking around Lipson's quiet lab, I sense something missing. "You have real students sitting at the computer benches. I don't see any chatbots."

Lipson smiles indulgently. His chatbots have been a YouTube craze. "That was just an afternoon hack. It went viral in twenty-four hours and took us completely by surprise."

He doesn't mean "hack" in its usual sense of breaking into a computer with malicious intent, but as highwire digital artistry. The *Urban Dictionary* defines its slang use like this: "v. To program a computer in a clever, virtuosic, and wizardly manner. Ordinary computer jockeys merely write programs; hacking is the domain of digital poets. Hacking is a subtle and arguably mystical art, equal parts wit and technical ability, that is rarely appreciated by non-hackers."

One day, Lipson asked two of his PhD students to bring a demo chatbot to his Artificial Intelligence class. Acting a bit like a portable, rudimentary psychotherapist, a chatbot is an online program that reflects what someone says in slightly different words and asks open-ended questions. It can come across as surprisingly lifelike (which says a lot about the clichés that pass for everyday chitchat). But in 1997 a "Cleverbot," designed by the British AI expert Rollo Carpenter, went online with a teeming arcade of phrases compiled from all of its past conversations. Each encounter had taught it more about how to interact with humans, including the subtleties of innuendo and pricks of friendly debate, and it learned to apply those nuances in the next chat. Since then it's held twenty million conversations, and its verbal larder is a treasury (or a snakepit) of useful topics, ripe phrases, witty responses, probing questions, defensive expressions, and the subtle rules of engagement, gleaned from years of bantering with humans.

Lipson's grad students set the laptops face-to-face on a table so that they could tête-à-tête in a virtual parlor. On one screen a computer-generated male materialized, on the other screen a female. The man spoke with a slight British accent, the woman in a syncopated Indian voice. Fortunately, the grad students videotaped the encounter and posted it online, where the chatty Cleverbots have now enchanted over four million people with their oddly human conversation.

The robots begin with a simple "Hello there," followed by pleasantries, but as they respond to one another they soon start to disagree, and the exchange grows funny, poignant, snarky, and thoroughly hypnotic.

"You were mistaken," Mr. Cleverbot says to Ms. Cleverbot, adding sarcastically, "which is odd, since memory shouldn't be a problem for you!"

"What is God to you?" she asks him at one point.

"Not everything," he says. It's a surprisingly plausible answer.

"Not everything could still be something," she insists with jesuitical aplomb.

"Very true," he concedes.

"I would like to believe it is."

"Do *you* believe in God?" he asks.

"Yes I do," she says emphatically.

"So you're Christian . . ."

"No I am *not!*" she snaps.

They bicker and make nice-nice. He calls her a "meanie," for not being helpful. She suddenly asks him a painful question, one any human might wonder about. Still, it's disquieting to hear.

"Don't you want to have a body?"

And then, surprisingly, like someone who has accepted a fate he nonetheless laments, he answers: "Sure."

What else is there to say? Abruptly they freeze into replica humans once more, and the video clip is over. Some people detect animosity or sexual tension between the man and woman, others a marital spat. We're ready to accept fictional robots in movies and stories, but are we ready for a synthetic life form that feels regret, introspects, and conducts relationships—creatures opaque to us, whose minds we can't fully mirror? Do the chatbots appeal because they're so like us, or because we're so like them?

There are scores of people in robotics who can fine-tune a robot's movements, even design truly lifelike robots with delicately mobile faces. Italian roboticists, for example, have created a series of realistic-looking heads that synchronize thirty-two motors hidden beneath the robots' polymer skin, and mimic all of our facial expressions, based on muscle movements, and can even capture the emotional space between furrowing the brows, say, and frowning. Such robots have already passed the stage of being a mere sensation in the robotics world. Fully-featured human faces are smiling, grimacing, exchanging knowing looks the world over. Unlike Madame Tussaud's wax-museum stars, today's robots look lifelike enough to seem a bit creepy, with facial expressions that actually elicit empathy and make your mirror neurons quiver. Equally realistic squishy bodies aren't far behind. One can easily imagine the day, famously foretold in the movies *Blade Runner* and *Alien*, when computers with faces feel silicon flavors of paranoia, love, melancholy, anger, and the other stirrings of our carbon hearts. Then the already lively debate about whether machines are conscious will really heat up. This was always the next step toward designing a self-aware, agile, reasoning, feeling, moody *other,* who may look like you or your sibling (but have better manners).

No doubt "robot sociology" and "robot psychology" will emerge as important disciplines, because there's an interesting thing that happens when robots become self-aware. Just like people, they sometimes get wrong impressions of themselves, skewed enough to create robot delinquents, and we might start to see traits parallel to psychological problems in humans.

When I used to volunteer as a telephone Crisis Line counselor, it wasn't always easy finding ways to help the callers who phoned in deep despair or creased by severe personality disorders. Self-aware robots with social crises, neuroses, even psychoses? That might prove a challenge. Would they identify with and prefer speaking to others of their kind? Suppose it concerned a relationship with humans? Colleges have popular schools of "International Labor Relations," "Human Ecology," and "Social Work." Can "Interspecies Labor Relations," "Robot Ecology," and "Silicon Social Work" be far behind? How about a relief order for aged, infirm, or incarcerated robots, such as "Android Daughters of Charity" or "Our Sisters of Perpetual Motion?"

What would the *Umwelt* (worldview encompassing thoughts, feelings, and sensations) of a self-aware robot be like? We're no longer entertaining such ideas merely as flights of imagination, but contemplating how to behave in a rapidly approaching future with the startling technology we're generating. If, as Lipson says, our new species of conscious, intelligent robots will learn through curiosity and experience, much as children do, then even robo-tots will need good parenting. Who will set those codes of behavior—individuals or society as a whole?

CAN WE LIVE inside a house that's a robotic butler, protector, and chatbot companion all rolled into one, an entity with its own personality and metabolism? Its brain would be a robotic Jeeves (or maybe Leaves), who tends the meadow walls and human family with equal pride, and is a good listener, with a bevy of facial expressions. A fully

butlered house with a face that rises from a plastic wall would monitor the energy grid, fuel the car (with hydrogen), while exchanging
news, ordering groceries, piloting a personal drone to the post office,
and preparing a Moosewood Restaurant lunch recipe that includes
herbs from the herb-garden island in the kitchen, and arugula and
tomatoes from the rooftop garden. In some high-tech enclaves, smart
locks are now opened by virtual keys on iPhones, and family members wear a computer tracking chip that stores their preferences. As
they move through each room, lights turn on ahead of them and
fade away behind, a thermostat adjusts itself, the song or TV show
or movie they were enjoying greets them, favorite food and drink
are proffered. The house's nervous system is what's known as the
"Internet of Things."

In 1999, the technology pioneer Kevin Ashton coined the term
for a cognitive web that unites a mob of physical and virtual digital devices—furnace, lights, water, computers, garage door, oven,
etc.—with the physical world, much as cells in the body communicate to coordinate actions. As they cabal among themselves, syn-
chronizing their energy use and activities, they can also share data
with the neighborhood, city, and wired world.

Combining animal, vegetable, mineral, and machine, his idea
is playing out in the avant-garde new city of Songdo, South Korea,
where the Internet of Things is nearly ubiquitous. Smart homes,
shops, and office buildings stream data continuously to a cadre of
computers that sense, scrutinize, and make decisions, monitoring
and piloting the whole synchronous city, mainly without human
help. They're able to analyze picayune details and make sure all the
infrastructure hums smoothly, changing traffic flow during rush
hour as needed, watering parks and market gardens, or promptly
removing garbage (which is sucked down through subterranean
warrens to a processing center where it's sorted, deodorized, and
recycled). Toiling invisibly in the background, the council of computers can organize massive subway repairs, or send you a personal
cell phone alert if your bus is running late.

It's a little odd thinking of computers taking meetings on the fly and gabbing together in an alien argot. But naming it the Internet of *Things* domesticates an idea that might otherwise frighten us. We know and enjoy the Internet, already older than many of its users, and familiar now as a pet. In an age where even orangutans Skype on iPads, what could be more humdrum than the all-purpose, nondescript word "things"? The Internet of Things reassures us that this isn't a revolutionary idea—though, in truth, it is—just an everyday technology linked to something vague and harmless sounding. It doesn't suggest brachiating from one reality to another; it just expands the idea of last century's cozy new technology, and animates the idea of home.

In J. G. Ballard's sci-fi short story "The Thousand Dreams of Stellavista," there are psycho-sensitive houses that can be driven to hysteria by their owners' neuroses. Picture sentient walls sweating with anxiety, a staircase keening when an occupant dies, roof seams fraying from a mild sense of neglect. Some days I swear I'm living in that house right now.

PRINTING A ROCKING
HORSE ON MARS

For centuries, the world's manufacturing has been a subtractive art, in which we created artifacts by cutting, drilling, chiseling, chopping, scraping, carving. As a technology, it's been both mind-blowing and life-changing, launching the Industrial Revolution, spawning the rise of great cities, spreading the market for farm-raised goods, and wowing us with everything from ballpoint pens to moonwalkers. It's still a wildly useful method, if sloppy; it creates heaps of waste and leftovers, which means extracting even more raw materials from the earth. Also, mass-produced items, whether clothing or electronics, require a predicament of cheap labor to add the final touches.

In contrast, there's "additive manufacturing," also known as 3D printing, a new way of making objects in which a special printer, given the digital blueprint for a physical item, can produce it in three dimensions. Solidly, in precise detail, many times, and with minimal overhead. The stuff of *Star Trek* "replicators" or wish-granting genies.

3D printing doesn't cut or remove anything. Following an electronic blueprint as if it were a musical score, a nozzle glides back and forth over a platform, depositing one microscopic drop after another in a molten fugue, layer upon layer until the desired object rises like a sphinx from the sands of disbelief. Aluminum, nylon, plastic, chocolate, carbon nanotubes, soot, polyester—the raw material doesn't matter, provided it's fluid, powder, or paste.

Hobbyists share their favorite digital blueprints via the Internet, and some designs are licensed by private companies. Like many other technologies, 3D printing does have a potential dark side. People have already printed out handguns, brass knuckles, and skeleton keys that can open most police handcuffs. Future laws will undoubtedly restrict access to illegal and patented blueprints, and also to dangerous metals and gases, explosives, weapons, and maybe the fixings for street drugs.

Imagine being able to press the print button whenever you want a candelabra, toothbrush, matching spoon, necklace, dog toy, keyboard, bike helmet, engagement ring, car rack, hostess gift, stealth aircraft rivets—or whatever else need or whim dictates. The Obama administration announced that it had seen the future and was investing $1 billion in 3D printing "to help revitalize American manufacturing." According to scientists and financial analysts alike, within a decade household 3D printers will be as common as TVs, microwaves, and laptops. However, people will still need to buy supplies and copyrighted blueprints for home printing, and many will order 3D objects ready-made from cottage industries.

In the future, even in the Mars colony Olivine calls home, she could fabricate a rocking horse of exactly the right height and dappled pattern on the morning of her daughter's birthday. Or she might print an urgently needed pump, and then a set of demitasse spoons with Art Deco stems. Or paint shades that don't yet exist in tubes. Artifacts that can't be created in any other way, such as a ball within a ball within a ball within a ball. Or an item with a hundred moving parts that's printed as a single piece. From this strange new forge, who knows what artworks and breakthroughs will emerge. The creative opportunities are legion.

We may ignore all the traditional limits set by conventional man-ufacturing. With micrometer-scale precision, we can seal materials within materials, and weave them into stuff with bizarre new struc-tural behaviors, like substances that expand laterally when you pull them longitudinally. A brave new world of objects.

What is an object if you can grow it in your living room drop by drop or molten coil upon coil? How will we value it? Today, because 3D printing is still a novelty for many people, we value its products highly, in wonderment. But when cheap home 3D printers become commonplace (today's cost anywhere from $400 to $10,000), and factory 3D printing replaces the assembly lines and warehouses, and even body parts and organs can be made to order, we'll live in an even more improbable world, where some objects continue to exist as tangible *things*, as merchandise, but a great many will exist concretely but in nonmaterial form, in a cloud or in a cartridge of fluid or powder, the way e-books do, as quickly accessible *potential*.

As cars, rockets, furniture, food, medicine, musical instruments, and much more become readily printable (some of those already are), it's bound to temporarily unnerve the world's economies. After all, we value things according to their scarcity. When gold is plen-tiful, it's cheaper. But if objects lurk as software codes, inside com-puters, and are abundantly available at the push of a button, they'll exist as another class of being. How will that change our idea of matter and the physical reality of all that surrounds us? Will it lead to an even more wasteful world? Will handcrafted objects become all the dearer? Will the Buddhist doctrine of nonattachment to worldly things flourish? Will we become more reckless?

This may all seem far-fetched, but not so long ago the Xerox machine was a leap of faith from carbon paper. When I first worked as a professor, making a carbon copy—what the "cc" on e-mail stands for—was a part of daily life. It's still somewhat astonishing to me that we can now print images in *color*, from *home* machines that can connect to our computers through the *air*.

Many companies won't look the same, because they won't need

to hire scores of workers, buy raw materials, ship or stock or *produce* anything. Industry, as we know it, may end. Financial advisers, business magazines, and online investment sites such as the Motley Fool believe 3D printing companies will clean up big-time, because their overhead will be so much lower, and they'll sell only the clever designs or raw materials.

Not right away. Most people will probably still find it more convenient to buy ready-made things. But soon enough, in the next fifteen years, 3D printing will revolutionize life from manufacturing to art, and practical visionaries like Lipson feel certain it will usher in the next great cultural and psychological revolution. For some, that future is the obvious sequel to the digital revolution. For others, it's as magical as a picture painted on water.

"Just like the Industrial Revolution, the assembly line, the advent of the internet and the Social Media phenomenon," *Forbes* magazine forecasts, "3D Printing will be a game changer."

How close are we to that day? It's already dawned. 3D printers are whipping up such diverse marvels as drone aircraft, designer chocolates, and the parts to build a moon outpost from lunar soil. Already, the TV host Jay Leno uses his personal 3D printer to mint hard-to-find parts for his collection of classic cars. The Smithsonian uses its 3D printer to build dinosaur bones. Cornell archaeologists used a 3D printer to reproduce ancient cuneiform tablets from Mesopotamia. Restorers at Harvard's Semitic Museum used their 3D printer to fill in the gaps of a lion artifact that was smashed three thousand years ago. In China's Forbidden City, researchers use a 3D printer to inexpensively restore damaged buildings and artworks. NASA used 3D printing to build a prototype of a two-man Space Exploration Vehicle (an oversized SUV astronauts can live in while they explore Mars). A USC professor, Behrokh Khoshnevis, has devised a method known as Contour Crafting for printing out an entire house, layer by layer—including the plumbing, wiring, and other infrastructure—in twenty hours. When 3D printers are linked to geological maps, houses can be made to fit their terrain perfectly. Khoshnevis is designing both single houses and colonies for urban planning, or for use after hurricanes, tornadoes, and other natural disasters when fully functional emergency houses will be 3D-printed from the ground up.

Boeing is 3D-printing seven hundred parts for its fleet of 747s; it's already installed twenty thousand such parts on military aircraft. The military's innovative design branch, DARPA, which began funding 3D printers two decades ago, finds them invaluable for repairing fighter jets in combat or supporting ground troops on the front lines. They're superb at coining parts instantly, remotely, to exact specifications, without having to wait for urgently needed supplies, or risk lives to ferry them through hostile terrain. Companies like Mercedes, Honda, Audi, and Lockheed Martin have been fashioning prototypes and creating numerous parts inside 3D printers for years. Audi plans on selling its first 3D-printed cars (modules printed then robot-assembled) in 2015.

The Swiss architect Michael Hansmeyer has 3D-printed the world's most complex architecture: nine-foot-tall Doric columns of breathtakingly intricate swirling organic laces, crystals, grilles, pyramids, webs, beehives, and ornaments, madly rippling around, fainting through, vaulting from, and imbedded into each other as layers of exquisitely organized chaos that began as a mirage in the mind and hardened. Containing sixteen million individual facets and weighing a ton, it looks like a roller-coaster ride down a scanning electron microscope into the crystalline spikes of amino acids. It's easy to imagine a cathedral by Antoni Gaudí with such columns in Barcelona. Or the labyrinthine short stories the Argentine fabulist Jorge Luis Borges might unleash among them.

"Twenty-five-year-olds today aren't burdened with traditional methods and rules," says Scott Summit, who heads Bespoke Innovations, a San Francisco–based firm that uses 3D printing to create elegant, tailor-made prosthetic devices. "There are guys who have been doing 3D modeling since they were eleven and are caffeinated and ready to go. They can start a product company in a week and, in general, have a whole new take on what manufacturing can be."

Since anything that can be designed on a computer and squirted through a nozzle is 3D-printable, people overwintering in Antarctica or other remote outposts will soon print their own cleaning products, medicines, and hydroponic greenhouses.

This blossoming technology widens the dream horizon of research, paving the way for new pharmaceuticals and new forms of matter. At the University of Glasgow, Lee Cronin and his team are perfecting a "chemputer," as well as a portable medicine cabinet so that NATO can disperse drugs to remote villages, especially simpler drugs such as ibuprofen. Despite unleashing an inner circus, most drugs are only a combination of oxygen, hydrogen, and carbon. With those simple inks and a supply of recipes, a 3D printer could concoct a sea of remedies. Flasks, tubes, or unique implements might also be printed on the spot. Creating new substances with 3D printers, researchers will be able to mix molecules together like a basket of ferrets and see how they interact. Then, as drug companies patent the recipes, those recipes (not the drugs) will hold value, just as apps do.

With 3D printers, complexity is free. For the first time, making something complicated with crisp details and ornate features is no harder than making a spoon or a paper weight. After the design component, it requires the same amount of resources and skills. That's a first in manufacturing, and a first in human history. If one person, regardless of skill or strength, can replace an entire factory, then identity and sense of volition are bound to shift. Will we all feel like kingpins of industry? No more so than most people do today, I imagine. But we should.

In research labs and medical centers all over the world, bioengineers are printing living tissue and body parts. That, too, is a first in human history, and a radical departure in how we relate to our bodies—not as fragile sacks of chemicals and irreplaceable organs, but as vehicles whose worn or damaged parts may be rebuilt.

In 2002, the bioengineer Makoto Nakamura noticed that the ink droplets deposited by his inkjet printer were about the same size as human cells. By 2008, he had adapted the technology to use living cells as ink. A regular 3D printer extrudes melted plastic, glass, powder, or metal and deposits the droplets in minuscule layers. More droplets follow, carefully placed on top of the previous ones in a specific pattern. The same is true for bioprinting, but using the patient's own cells reduces the chance of rejection. Each drop of ink contains a cluster of tens of thousands of cells, which fuse into a shared purpose. Although one can't control the details, one doesn't need to, because living cells by their fundamental nature organize themselves into more complex tissue structures. The hope is to be able to repair any damaged organ in the body. No more worrying about size or rejection, no more waiting for a kidney or liver to become available.

Today, in university and corporate labs around the world, bio-engineers are busily printing ersatz blood vessels, nerves, muscles, bladders, heart valves and other cardiac tissues, corneas, jaws, hip implants, nose implants, vertebrae, skin that can be printed directly onto burns and wounds, windpipes, capillaries (made elastic by pulses from high-energy lasers), and mini-organs for drug testing (bypassing the need for animal trials). An Italian surgeon recently transplanted a bespoke windpipe into a patient. Washington State University researchers have printed tailor-made human bones for use in orthopedic procedures. An eighty-three-year-old woman, suffering from a chronic infection in her entire lower jaw, had it replaced with a custom-built 3D titanium jaw, complete with all the right grooves and dimples to speed nerve and muscle attachment. Already speaking with it in post-op, she went home four days later.

A team of European scientists has even grown a miniature brain for drug tests (though, fortunately, it's not capable of thought). Organovo, a leading biotech company in San Diego, has 3D-printed working blood vessels and brain tissue, and successfully transplanted them into rats. Human trials begin soon. After that, Organovo plans to provide 3D-printed tissues for heart bypass surgery. Meanwhile, a kidney is the first whole organ they're working on—because it's a relatively simple structure.

Thin body parts like these are the easiest to design. Thicker organs, such as hearts and livers, require a stronger frame. For that, a lattice of sugar—like the haute cuisine sugar cages some chefs confect for desserts—is often used to provide a firm scaffolding, and then cells are layered over it. Sugar is nontoxic and melts in water, so when the organ is finished, the sugar scaffold is rinsed away, leaving hollow vessels for blood flow where they're needed. The goal isn't to create an exact replica of a human heart, lung, or kidney—which after all took millions of years to evolve—nor does it need to be. A kidney cleans the toxins from the blood, but it doesn't have to look like a kidney bean or a kidney-shaped swimming pool. So it could become body art, a sort of interior tattoo: a heart-shaped kidney for a romantic, a football-shaped one for a sports fan. Or would that alter the brain's mental atlas of the body, a landscape we know by heart, even in the dark? Suppose you have a suitcase. You replace the handle, you replace the lock, you replace the panels. Is it the same suitcase? If we replace enough body parts, or don't choose exact replicas, will our brain still recognize us as the same self?

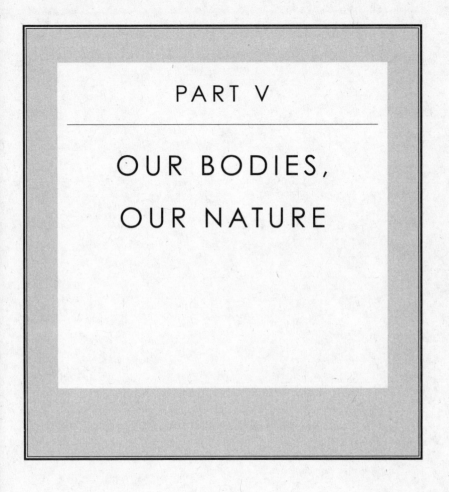

PART V

OUR BODIES,
OUR NATURE

THE (3D-PRINTED) EAR
HE LENDS ME

L awrence Bonassar's lab in Cornell University's Weill Hall sits
across the street from a jewel of a tiny flower garden, now blanketed in snow. Though Weill's outside walls are white as the
season, it's one of the "greenest" buildings in the country, which has
earned a rare gold LEED rating, thanks to everything from recycled
building debris and materials (such as the outer skin's white aluminum panels) to a cooling roof planted with succulents and flowers,
heat-reflective sidewalks, a giant atrium with passive solar heat, and
motion sensors that turn on lights, temperature, and air flow only as
needed, when people appear.

Opened five years ago, as a state-of-the-art home for the Department of Life Sciences, it's been designed as an intellectual crucible
with large overlapping labs. Long open sunlit rooms, running the
length of each floor, share common lounges, corridors, and microscope areas, making it impossible *not* to bump into postdocs in
related fields. Even in winter, cross-pollination is encouraged. Just
as the planners hoped, many collaborations have ensued, the new

field of regenerative medicine is taking wing, and bioprintmakers are crafting tailor-made body parts.

The principle of "regenerative medicine" is magically simple: if a heart or jaw is damaged, either teach the body to regrow another or print a healthy new one the body will embrace. What Bonassar's lab regenerates is the body's vital infrastructure of cartilage: all those cushions (the so-called disks) between the vertebrae in the spine; the easily torn meniscus in the knee; the accordion of semicircular rings that keeps the trachea from collapsing when we breathe, yet allows it to bend forward when we swallow; the external ear, prized by poets and nibblesome lovers, who often describe it as "shell-like." His goal is to restore lost function to ragged physiques and fix facial defects. To that end, he mingles tools from several disciplines, including biomechanics, biomaterials, cell biology, medicine, biochemistry, robotics, and 3D printing. If your only tool is a ruler, you'll tend to draw boxes. New tools create new mental playgrounds. On this playground, spare ears abound.

What could you do with a spare ear? If a patient with skin cancer has to have an ear removed, the traditional way to replace it is with a prosthetic, but that has to be donned every day. A young mother in exactly that fix lamented to CBS News, "I could just see my kids running around with it, yelling 'I have mommy's ear!'" Instead, surgeons at Johns Hopkins harvested cartilage from her ribs, shaped it into an ear, and implanted it under her forearm skin, where it could be nourished by her own blood vessels. Just four months later, when the ear had grown its own skin, it was removed from her arm and attached to her head. Yet, wonderful as her new homegrown ear was, the process required numerous operations, including breaking open the vault of the chest and stripping cartilage from the ribs, then shaving and shaping the ear to fit. A feat of *subtractive* manufacturing.

Or an ear might rescue the one child out of nine thousand who is born with microtia, a condition in which the external ear hasn't fully developed, sometimes leaving only a small peanut-shaped vestige behind for classmates to ogle and mock. A father of three young children—twin six-year-old girls and a five-year-old boy—Bonassar is deeply sympathetic to the condition, and mindful of what early fixes can mean. Unfortunately, children aren't able to brave an operation like the young mother's until they're six to ten years old, because they don't have enough rib cartilage. Also, the operation is very painful and quite traumatic, and you don't really want to subject a child to it. How much simpler, less invasive, and cheaper to do an MRI, CT scan, or 3D photograph, then print the cartilage out on demand in the exact shape of each child's highly personal ear. You'd be able to do it much earlier in a child's life, and you could photograph the left ear, flip it around to make a right ear, and match the geometry perfectly.

Microtia doesn't harm hearing, but it often invites the social nightmares of bullying and shunning, just as a child is detailing a sense of self. So, although a new ear is only a cosmetic change, it has an enormous impact on a child's hope of making friends—which in turn shapes a growing child's brain. The ability to smile is a child's coin of the realm, pleasant ears and face its passport. As I learned volunteering for a short spell with Interplast in Central America, the sooner you can repair a harelip, birthmark, or other deformity, the better chance a child has to bond with her parents, let alone strangers.

I don't really need a new ear. Except for the occasional snaredrum of tinnitus or missed stage whispers, mine are working passably well. And the outer shells fit the size of my head. I could use more cartilage in my left knee, and a new spinal disk one day, but I don't fancy today's remedies—a cold metal implant, or a gift from a four-legged animal.

Nonetheless, a homegrown ear is what Lawrence Bonassar offers me, extending it in his open hand, as if it were sprouting out of his palm and shaking hands were just another form of listening. Translucent white, the ear feels smooth and warm as amber, and my thumb automatically plays over its many ridges and folds. I'm surprised by the level of minute detail. It's quite odd to fondle a disembodied ear. Or a prizewinning one, for that matter. His bioprinting has won first place in the World Technology Awards in Health and Medicine, the Oscars of the technology world, which celebrate inventions "of the greatest likely long-term significance for the twenty-first century."

The ear he lends me is solid, yet bendy as a dried apricot, and would flex easily under the skin. But this one isn't intended for anyone's head. He returns it to a glass jar of preservative and sets it back on a shelf. His lab looks like the mental crossroads it is: a chemistry lab full of microscopes, workbenches, sinks, glass, and stainless steel. But also a medical facility, complete with large incubators, sterile fields, and countless drawers of parts and molds. And also a tech center equipped with computers, robots, and of course 3D printers. All accompanied by a seemingly endless wall of windows.

Outside there is the frail enchantment of snow, and a school bus creeping by like the orange pupa of some colorful butterfly.

One long ray of sunlight like a pointing finger touches a white desktop box about the size of the first manual typewriter, but less complicated looking. Two steel syringes with nozzles float above a metal warming plate, where they'll begin hope's calligraphy. The ink may be living cells of any type, life's pageant in a polymer. When the nineteenth-century painter Georges Seurat used a similar stipple technique with pure color, his dots seemed to blend, but that was merely sleight of eye. These dots merge because cells fuse freely; they don't need a human nudge.

As the moving pen writes, it doubles back over each line, stippling new layers, until it creates an outer ear that isn't exactly organic in the way a wart or an eyelash is. Still, when transplanted, it will twitch with life, feel embodied, and help redefine what we mean by "natural." Then any tangle of flesh and blood will serve as home, making rejection obsolete. For once in our long-storied evolution, a body part isn't molded solely by evolution's blueprint—we can choose its design. And forget years. Once he receives the MRI, CT, or 3D image, Bonassar can grow an ear in fifteen minutes. The time it takes me to walk to my local coffee shop.

Bonassar has mastered the art of training materials to carry cells and deposit them like puppies exactly where he wants while keeping them alive and happy. And, just like healthy pups, the cells are agile enough to tussle without smooshing, hale enough to tug, eager to curve their tiny mouths inward and eat all the nutrients they need.

The two polymers he prefers are collagen—protein fibers the body uses as gluey twine and mortar—and alginate, a gel found in the brown seaweed I held at Thimble Islands, and used by drive-ins to make milk shakes thick and creamy as they swirl out of the dispenser. My parents had one of the first McDonald's, where I sometimes poured shakes, and I know the consistency well—between syrup and toothpaste. That's also the way bioprinters dispense ink, except that the carrier has clusters of living cells embedded in it.

When I ask Bonassar if he makes the scaffolds here, too, a proud smile appears, as he explains that the scaffolding *is* the liquid. This shake contains its own framework, so that an ear fresh from the printer is ready to go. And the ink? Water droplets wouldn't stick together. Hard marbles would roll off. Instead he uses squishy collagen marbles, which cling to their neighbors. Like eggs or blood, collagen gels when it's heated and the fibers scramble together. So he stores it cold, allowing it to stiffen only when it falls onto the warming plate.

It's a technology he hijacked from Hod Lipson, and he began printing in Lipson's lab with a printer wide as a brick hearth and heavy as an iron cauldron. Now it's the size of an espresso machine and shockingly simple: load whatever you want into your syringe, place it so that the motor can grab the plunger and print, then adjust the rate ink squeezes out, and set the print head's path. After that, two more steps remain.

He leads me through an open doorway into a tiny room packed with large machines, including a sterile culture hood with a large window, in which the whole bioprinter can be placed, safe from dust, fungi, and bacteria. It looks like an incubator for preemies.

"No, *this* is the incubator," he says, turning to peer through a glass door into dimly lit shelves. "Look there. Can you see them?"

Rising on tiptoes, I see two petri dishes holding small strange buttons or perhaps tires. The odd items are spinal implants destined for a bouncing, braying fluff of a dog. Canine arthritis is a big problem for frolicsome dachshunds, hounds, and especially beagles—dogs with short necks and long bodies wear out their cervical disks and develop joint pains just as we do. Working with Cornell's Veterinary School, Bonassar's lab created implants with a gel-like core that pushes against a tougher outside ring, pressurizing it very much like blowing up a tire. It's also a true organ, two different kinds of tissue that work together seamlessly.

In osteoarthritis, cartilage's cushion wears out like an old pillow, and bone rubs bone raw, producing inflammation and pain. Almost everyone has a creaky-jointed sufferer in the family. Today's back operations usually remove a damaged disk and fuse the vertebrae together with a metal plate, which creates the rigidity of a poker up the spine, doesn't always work, and can make adjoining vertebrae weak as loose teeth. An alternative to fusion would be a godsend to sixty or seventy million people in the United States alone. Starting small, Bonassar replaced spinal cartilage in rats with his own lab-grown variety, and the rats lived normal lives, apparently pain-free. Next in line are larger animals—a dog, sheep, or goat—and if that works, then human volunteers will follow.

Bending to examine a smaller chamber alongside the incubator, I spot the next stage in the life of bespoke disks. Since all tissues in the body are weight-bearing and thrive under stress, his lab toughens the tissues, squeezing the implants over and over, as if they were pumping iron. This also quickens their metabolism, squooshing food in and smooshing waste out, making it more efficient. Such bioprinted implants could last longer than natural ones.

"It's quite realistic to assume," Bonassar says, hazel eyes sparkling, "that the first stages of the human clinical trials could happen within the next five years."

When I ask about printing out hearts, lungs, and livers, he leads me to a large computer screen where he summons up a pair of gloved hands holding what look like pieces of sushi: thick white slices with a thin halo of pink. A closer look reveals a rarer delicacy: ear implants fabricated from 3D photographs of Bonassar's daughters' ears. He smiles at them with a love pure as starlight. Then he points out the blood vessels in the thin rind around the implants, and the thicker comma of white tissue that's quite bare. Yet those bare cells prosper, too.

The challenge with organs isn't their size, it's the plumbing. The bigger organs are like Venetian cities, fed by elaborate water streets afloat with gondolas. Many labs around the world are hunting the best ways to mirror those supply lines, and the elusive "aha moment" could be one week away or ten years. But its scent is in the air, and no one doubts it will soon revolutionize medicine with clean, healthy organs on demand. A touch of mental whiplash is to be expected.

It's a hallmark of the Anthropocene that science and technology are galloping at such a pace that Bonassar's field didn't even exist when he was in high school or college. Now he's among those ringing the biggest changes, including a dramatically new view of the human body and the jostles of cells that inhabit living tissues—even what a cell is and how a cell behaves. Not only do we know about stem cells, we're starting to wield them in clever ways to mend the body, and it's not arduous to do; it can be as simple as exposing cells to the right chemicals or stimuli. There's been a stunning paradigm shift from the rule of *phenotype*—one cell type fated for one job and nothing else—to *phenotypic elasticity*, the idea that cells are far more versatile and can be repurposed, like a hammer used to anchor a kite.

We now grasp that a wafer of skin can be retrained to do just about anything. It's a new category of raw material, like wood or stone, with potent gifts. An ebony tree growing in Africa may provide shade to humans, and a lofty haven for a leopard gnawing a carcass, but its dark grain also gives rise to clarinets, piano keys, violin fingerboards, and music. The old idea of skin as a sacred cloak with two main jobs—to seal off the vulnerable organs inside us and define our individuality—has given way to a sense of how mingling, malleable, and porous the body really is. At the cellular level, we're stunningly mutable, not just in our lifestyles, which we always knew, but in our bits and pieces. A butler can change his mind, via his neurons, and become a gandy dancer. A dash of skin can become fresh neurons for a Parkinson's-stricken monk. What to do with cells is increasingly more a question of imagination than material.

Spearheaded by pioneers like Bonassar and Lipson, Anthropocene engineering has penetrated the world of medicine and biology, revolutionizing how we regard the body. In these vistas, electricity, architecture, and chemistry slant together and tell tales never heard before.

We baby boomers grew up with a cartload of absolutes, handed down from generation to generation of biologists, the most daunting one, perhaps, being that we're born with all the brain cells we'll ever have, because the brain doesn't mint new cells. Yet now we have proof that it does, even in old age. We've spent the last decade blowing up a lot of similar assumptions, and I wonder what other rigid ideas will topple. Bonassar offers me another quite mysterious one.

"We were told, over and over," he says with relish, "that the heart is an organ that positively can't regenerate. Yet an amazing study has turned that idea upside down."

In this study, he explains, heart transplant patients received hearts from a donor of a different gender—mostly men receiving women's hearts. In theory, one should be able to examine the heart recipient, look at the cells in his borrowed heart, and find female cells from the donor. But it turns out that, on autopsy, if these men bore their transplanted hearts for more than a decade, almost half of the heart cells were mysteriously replaced by male cells. The mechanism isn't clear—but the new paradigm is. The heart's metronome tissues, which we always believed couldn't regenerate, actually can. No one knows if the cell-swap is a fusion or whether the female cells were forcibly displaced. Either way, it's overturned the handcart of possibility, and furrowed many brows. If organs as elemental as brain and heart can be persuaded to regenerate, and others, like ears and corneas, can be fashioned from living ink, how will that change us as a species? Will the printing of organs affect our evolution? Could it alter our genes? I'm curious to know what Bonassar thinks.

The possibility intrigues him, too. "The real question," he says, "is what the evolutionary pressure of these therapies might be. Would faulty genes become more prevalent, because they could be fixed? I wear contact lenses, but if I were a caveman and my eyesight was as bad as it was when I was five, I would have been in serious trouble. Now it doesn't matter. We could replace our defective parts, live longer, and feel healthier." Then he adds a provocative after-thought: "Yet physically we could be much weaker and more flawed genetically."

Suppose we don't just repair and enhance ourselves, suppose we live longer as a result? The primary focus of the work in Bonassar's lab—cartilage for arthritis, cartilage for traumatic injury, disks for back pain—is medical solutions for ailments that tend to afflict people long after they've had children, when evolution has stopped bedeviling them to breed. Would the ability to be fit for a decade longer present an evolutionary advantage? Would people take more risks? How will we regard the body's bits and pieces, and safeguard them, if we know we can cheaply replace them? We replace heart valves or heart tissue to extend life, but what if you can cure arthritis, and keep people active and sexual well into their seventies and eighties?

Think geriatric cyborgs and chimeras. Grandpa's going to be saying a lot more interesting things than *Where are my teeth!* Just staying active for an added decade may alter our society as a whole far more than fixing a particular defect in the heart, liver, brain, or kidney.

The ninety-year-olds I've known haven't run marathons, even with gleaming new hips and knees, but they've inhaled a lot of sky on daily walks. Even bioprinted cartilage needs exercise. Looking forward to a walkabout through drifting avenues of snow, I say good-bye to Bonassar and slip into my parka. As I stride down the hallway and into the atrium, the building's smart sensors work and a little breeze runs before me like an invisible serpent. Hail begins lightly rattling against the windows. A vague thought, as elusive as the smell of violets, nags at me. A dark cloud passes over, and I feel aware of how aging, like winter weather, can chill the bones. For a moment, that thought hangs like an icicle, tapering and cold. Then my mind reels through hopeful images: an incubator full of spinal disks, the flexible necks of dachshunds, the long open labs of students with eager minds, the children with new ears, and the warming plate where collagen marbles land on their way to reshaping our future biology, and I swear I hear spring buzzing like a red-winged blackbird.

CYBORGS AND CHIMERAS

At no point in my conversation with Bonassar did we discuss if people will *mind* the idea of artificial body parts. No need to. They're already a commonplace feature of the new normal. Not long ago the idea of a cyborg was pure science fiction, and we couldn't get enough of the Six Million Dollar Man (who inspired many a roboticist), *Star Trek*'s Captain Picard (who has an artificial heart), or the species of moody Replicants in *Blade Runner*. Now we think nothing of strolling around with stainless steel knees and hips; battery-operated pacemakers and insulin pumps; plastic stents; TENS pain units that disrupt pain signaling in the nerves; cochlear implants to restore hearing; neural implants for cerebral palsy, Parkinson's, or damaged retinas; polymer and metal alloy teeth; vaccines hatched in eggs; chemically altered personalities; and, of course, artificial limbs. A great many of us are bionic (I have a 5 cm titanium screw in my foot), and bionic hands, arms, legs, skin, hearts, livers, kidneys, lungs, ears, and eyes are all available. Visible cyborgs who move among us may grab our attention and curiosity, but they don't scare us anymore, and they're becoming commonplace.

On a windy November day in 2012, software designer Zac Vawter climbed 103 floors of Chicago's Willis Tower, the tallest building in the western hemisphere. From its Skydeck, 1,353 feet in the air, one can view four states and the pounded blue metal of Lake Michigan fanned out below. Breathing hard toward the end, with the 2,100th step he reached the Skydeck and strode straight into history.

It was a climb that challenged the stamina and knees of all 2,700 people who joined him to help raise money for the Rehabilitation Institute of Chicago. But what makes Vawter remarkable is that he did it using a gleaming new bionic leg. Surpassing even that is how he did it—by controlling the device with his thoughts.

A thirty-one-year-old father of two, Vawter lost his right leg in a motorcycle accident in 2009. Afterward he went through a pioneering procedure in which the residual nerves that once ruled his lower leg were "reassigned"—they were rerouted to control his hamstring. For months he flew to Chicago to work with engineers, therapists, and doctors to adjust the bionics and refine both his physical and mental technique.

As he pictured himself climbing—lifting his leg, bending his knee, flexing his ankle—electrical impulses from his brain flashed to his hamstring, which signaled a deftly designed assembly of motors, belts, and chains to lift his ankle and knee in unison, and he began taking the stairs step-over-step in the normal way. Just focusing hard doesn't work; he had to intend to walk. The bionic leg is designed to read an owner's intent, whether he's walking, standing, or sitting. So if he's seated and wants to stand up, he just pushes down and the leg pushes back, propelling him up.

Like any athlete, he had to prepare for months, while scientists tailor-made the prototype leg and he practiced the mind-feel of stair-climbing. In time, the brain accepted the robotics as an extension of his body image and took it into account when judging, say, whether or not he might fit through an open door. Yet when the climb was over and he flew home to Washington, he had to leave the leg in Chicago for researchers to continue tinkering with until it's even more reliable. Bionic arms are already popular, and if an arm fails someone might drop a glass of milk, or, more alarmingly, a baby, or a flaming match. If a bionic leg fails, they could tumble down a flight of stairs. So the technology has to be safe. RIC expects the FDA to approve such bionic legs within the next five years.

As long as humans have walked the Earth, we've been driven by a need to stretch into the environment; tools and technology have always been an innate part of that quest. Now we're comfortable with, and excited by, the promise of connecting our brains to the world outside of the body. iPads and cell phones that store phone numbers, calendars, to-do lists, photos, documents, and memories for us—external brains the size of a notepad—are just the beginning. *Oh, where did I leave my memories?* Most of us tuck our prosthetic memories in pockets, purses, and briefcases. On campus, the students tote spare hippocampi in their backpacks. We may fear losing our memory as we age, but at any age we're anxious about losing our prosthetic memories. Many people aren't at ease without obsessively "checking"—a verb once applied to OCD behavior (Did I turn off the stove? Close the garage? Shut the door tightly?). Relentless digital "checking behavior" has joined the closet of neurotic compulsions, and we've added these phobias to our quiver: *nomophobia/ mobophobia* (the fear of leaving your cell phone at home), *phantom vibrations* (thinking your cell phone is vibrating even though no one is calling), and *FOMO* (fear of missing out, and so relentlessly checking Facebook). *Continuous partial attention* (focusing halfheartedly) has become pervasive as we're tugged at by ringtones, text-tunes, incoming-mail pings, calendar flags, update alerts, new-post beeps, pop-ups, and the nagging possibility that something more engrossing may appear.

Our ancestors adapted to nature according to the limits of their senses. But over the eons, we've been extending our senses through visionary and stylish inventions—language, writing, books, tools, telescopes, telephones, eyeglasses, cars, planes, rocket ships—and, in the process, we've redefined how we engage the world but also how we think of ourselves. This even extends to our metaphors. We used to picture the body as a factory. Today that's sea-changed and scientists picture factories as primitive forms of cells. We used to compare the brain to a computer. Now DARPA has a SyNAPSE program whose goal is building "a new kind of computer with similar form and function to the mammalian brain."

Our cells dance with their own electric, and as they're immersed in ambient networks and signals—the *everyware*—we're becoming part of an invisible weave that's different from the one we used to picture as the seamless web of nature. This is part of the new natural. It slips beneath our radar for things weird, experimental, non-human. Anthropocene humans can merge with technology and not be regarded as alien.

Not only humans. When a puppy called Naki'o fell asleep in a puddle on a cold Nebraska night, he woke with frostbite on all four paws. As his condition worsened, Orthopets, a Denver company that specializes in prosthetics for animals, turned Naki'o into the world's first bionic dog. Equipped with four prosthetic limbs, he runs and romps normally with his owner—despite not being able to feel the ground—and has become the spokesdog for Orthopets, which has also equipped a front-legless Chihuahua with two wheels (he's a big hit in nursing homes). Other bionic animals include a wounded bald eagle, found starving in an Alaska landfill and given a new upper beak; a dolphin mutilated in a crab trap and unable to steer until it got a prosthetic tail; a green sea turtle with replacement flippers after a surfboard injury; and a baby orangutan, born with clubfeet, fitted successfully with therapy braces.

Clearly, these attachments were all prosthetics. But should we consider the first spears to be tools or imaginary prosthetics? Attacked by slashing bears, tigers, and other beasts with razory teeth and claws, our ancestors fought back by crafting teeth and claws of their own, ones they could detach and hurl from a distance. What a novel idea! Imagine a wolf flinging its teeth, one by one, at its prey. Stone axes were tools, but also prosthetic hands built stronger and bigger and sharper than a hominid's own. The first clothing was also a prosthetic, an artificial body part: skin. When cavemen and -women wore draped animal hides for warmth, tying them with sinew, no matter how much they tanned the clothing first, it would have yielded a whiff of other creatures. They slept under borrowed scents, in a cave permeated by sweet gamy odors that mingled with each person's personal bouquet. Today we're so far from the origins of our clothing that it's become impersonal, and we don't feel the powerful magic of being enrobed in another animal's skin or a plant's fibers.

Early humans probably devised crutches for the lame, but the first prosthetics are spoken of in the ancient sacred Indian poem the *Rig Veda*, where they belonged to the warrior queen Vishpla. After she lost a leg in battle, she still insisted on fighting, so some sort of iron leg was fashioned for her. The Greek historian Herodotus tells of a shackled Persian soldier who escaped by cutting off his foot and replacing it with a copper and wooden one. In the Cairo Museum, there's a mummy from the reign of Amenhotep II (fifteenth century BC), whose big toe on the right foot was amputated and replaced by a superbly carved wooden replica tied on with leather straps. She's believed to have been a royal woman suffering from diabetes, and the artful toe was designed to help her both on Earth and in the afterlife.

Throughout history, peg legs and hook hands have been plentiful, though such antique prosthetics were crudely made and heavy, usually from wood, metal, and leather. Wearing them, a person became part tree, part animal. We'll never know if kin regarded them as a hybrid, or if the wearers identified at all with the qualities of the species they harnessed in lieu of human muscle and bone.

How far we've come! Today we live in a completely prosthetic culture brimming with contact lenses, false teeth, hearing aids, artificial knees and hips, compasses, cameras, and many digital and wireless brain attachments. We've made such prosthetic strides since toes of leather and wood that it's even hard to agree on a fair playing field, since the ultimate Olympic athlete may be competing against a cyborg now. But is that fair?

Already a Paralympic gold medalist, Oscar Pistorius spent four years battling the Court of Arbitration for Sport for a chance to race against able-bodied athletes in the regular Olympics. Ultimately, after extensive testing of his blades, the court decided in his favor, declaring that the blades wouldn't give him an unfair advantage. Yes, the springy blades were lighter, but also limited; they couldn't return more force than Oscar generated striking the ground. In contrast, the elastic dynamo of a human foot and ankle can always pound with extra force and rebound with more velocity. So, at the moment, until the technology changes, able-bodied sprinters supposedly have an advantage over blade-wearing ones.

But doesn't every gifted athlete have some unique physiological advantage? For the swimmer Michael Phelps, the most decorated Olympic athlete of all time, it's an unusually long torso and arms for his height. The debate will heat up even more as higher-tech blades are invented. In an ironic twist, when Pistorius was outpaced by a sprinter in the Paralympics, he lodged a formal complaint that the winner had had an unfair advantage because he wore better-crafted blades.

Pistorius was the first double amputee ever to compete against able-bodied runners in the Olympics, and his story is a sort of double haunting, in which our past and future ghost into view. He is visibly a cyborg, and yet completely at home in his body. As a child, he fused mentally with his artificial legs, and his brain pictured blades as the natural extension of thighs, and his body as agile and fleet-footed.

Pistorius isn't the only famous cyborg. Wartime often leads to advances in technology, and as a result of all the young amputees returning from the Iraq and Afghanistan wars, the field of prosthetics has flourished, with high-tech materials and more natural-looking robotics. DARPA runs a Revolutionizing Prosthetics program, whose goal is an array of thought-controlled limbs that move with the precision and ease of natural limbs, ready for FDA approval in the next few years.

On the evening of November 6, when all the votes were counted for the 2012 election, Tammy Duckworth, an Iraq War veteran who had lost both of her legs on the battlefield in 2004, strode to the podium to make her acceptance speech as the newly elected Democratic congresswoman from Illinois. She wore state-of-the-art prosthetic legs complete with robotic, computer-controlled ankle joints and a computer-powered knee.

Using a cane, she moved smoothly and looked understandably elated, comfortable in motion, which is remarkable since Duckworth didn't grow up learning to balance her pelvis and spine over prostheses while she walked. As an infant she learned to walk fearlessly, as babies instinctively do, around the age of thirteen months, when balance and strength are keen enough, and some baby fat has yielded to muscle.

Although walking ultimately becomes unconscious, it's a skill that requires us to tumble into and out of balance all the time. It takes countless hours of practice and encouragement, and a great many falls to do it expertly. Babies are like tiny petulant stilt-walkers. To walk, you step forward with one foot, which tilts you off balance, then you catch yourself before you fall too far, quickly rebalance, and fall in the opposite direction, catch yourself, and start falling again as you make a so-called straight line across the room or street. Walking is really a series of recovered falls. In time we learn to do it expertly, without noticing that it's an evolutionary circus act. Over time, a lovely pendulum swing develops, as the hips roll out of balance and back in again, over and over, without the walker paying it any mind. Its rhythm is naturally iambic (a short unstressed syllable followed by a long stressed one), which could be why so many poets, from Shakespeare to Wordsworth, wrote poems in iambs; maybe they composed while strolling. Fortunately, the pelvis and backbone are engineered to make the skill (strolling, not composing poetry) relatively easy.

However, relearning to walk as an adult means unlearning old balancing tricks and mastering new ones, based on the current shape of your body, while *fully aware* that you could fall and badly injure yourself. Also, injuries aren't always symmetrical—Duckworth's were complex (right leg missing at the hip, left leg below the knee). Blades like Pistorius's wouldn't have suited her lifestyle, in which she needed to feel equally comfortable on airplanes, behind a desk, at a podium, or climbing stairs. Her revolutionary ankle joints and legs rely on robotic software—microprocessors, accelerometers, gyroscopes, and torque angle sensors—to mimic the delicate teamwork of muscles and tendons in the ankle when someone walks.

The day of the cyborg has certainly arrived, with goggles for skiers and snowboarders offering a dashboard display of data, GPS, camera, speedometer, altimeter, and Bluetooth phone; plus voice control and gaze control for cyclists. Will it be safe to ski or bike with data dancing before your eyes, or while posting photos on Facebook? Probably not. But safer than looking down at a smartphone and back up at the slopes, while dizzily jockeying between the sensory and tech worlds. The virtual reality that *Star Trek* promised us is starting to become commonplace. We can don a headset that stimulates all five senses simultaneously, and walk along a street in ancient Rome or Egypt, so immersed in the look, smell, and feel of the place that it seems real.

I've yet to meet anyone sporting Google Glass, the voice-controlled miniature screen in a flexible frame that hovers piratically above one eye, projecting e-mail and maps onto your visual field. But, whether or not it catches on as techno-fashion, it's already a triumph in operating rooms around the world. The first surgeon who wore it simply videotaped an operation to share with colleagues. Since then, surgeons have been actively consulting Glass during operations to view X-rays or medical data without turning away to look. The cyborg doctor has eyes in the back of his head, and four or more hands. At the University of Alabama at Birmingham (UAB), Dr. Brent Ponce, wearing a Google Glass, began a shoulder replacement surgery while the built-in camera showed the surgical field to Dr. Phani Dantuluri, a veteran surgeon watching on his computer monitor in Atlanta. As the doctors discussed the case, Dantuluri could reach into the surgical field that Ponce saw on his heads-up display: ghostly hands floated over the body, pinpointed an anatomical feature or demonstrated how to reposition an instrument, as he consulted in real time. Invented by a UAB neurosurgeon, Barton Guthrie, who was frustrated by the limits of teleconferencing, VIPAAR (Virtual Interactive Presence in Augmented Reality) offers a safety net in diverse situations: teaching surgeons, guiding a resident's hands, piloting difficult procedures in regional hospitals anywhere in the world, and also assisting emergency operations at an Antarctic base or in space.

A lively pair of glasses, even with all the digital trimmings, is still only an accessory. At the end of the day, you remove it and become mortal again. We yearn to supersize and supervise our powers with no intermediary, but intimately, naturally, without fuss, as if such marvels were our birthright. The next small step, a world closer, will be thought-controlled contact lenses floating on the eyes like high-tech continents. Will they cling invisibly, or twinkle like the glass eyes of a doll? The final step, who knows when, will be the silk of silicon sliding along our neurons. Then, without disappearing, only virtually visible, our computer worlds will fully mesh with us. Will we feel haunted sometimes, or merely worry about affording the latest update?

It's a strange paradox to imagine, yet highly possible, maybe even inevitable, but by delegating more physical and mental tasks to robots and computers, we might also weaken various skills and aptitudes—math, musculature, memory—while perfecting new ones. We may soon have to master multitasking spatially, as we cross midtown streets while scrolling through and answering e-mails hovering in the air, which we've conjured up on our iGlasses, just by batting our eyelashes. In time, brain and body would adapt.

We've always crafted new technologies to help us live better or longer, but in the past few decades that's accelerated dramatically.

We've stepped up the pace of our romance with machines, wedding them to our bodies as never before, and saturating our lives with techno-marvels, from genetic tests to organ transplants, satellite communications to genetic engineering, brain scans to mood enhancers. They've amused and nettled our lives to such an extent that a new branch of anthropology has arisen to study the phenomenon.

Amber Case practices "cyborg anthropology," a field in which scientists study how both humans and robots interact with objects, and how that changes the culture in which we live. For example, the way cell phones affect human relationships, and how we now interact techno-socially instead of socially. Old-fashioned social relationships, in which one gets together with friends, are regarded as "analog."

"So, for instance," Case explains, "we have these things in our pockets that cry, and we have to pick them up and soothe them back to sleep, and then we have to feed them every night by plugging them into the wall, right? And at no other time in history have we had these really strange nonhuman devices that we take care of as if they are real."

Cyborgs may be growing plentiful, but even more of us are chimeras—DNA (and sometimes body parts) from two or more creatures lodged inside one body. The ease with which mythic humans and animals breed and swap bodies speaks to our prior intimacy with the rest of nature, acknowledging animals as part of our extended family. We've adopted the term "chimera" from the Greek monstrosity Homer sang of in the *Iliad*, a savage sky-beast said to be part lion, goat, and serpent, that blasted fire from her mouth and terrorized the land until a hero named Bellerophon chased her on his winged horse, Pegasus, and rose above her fiery blasts to kill her. It's the sort of fiend that haunts many cultures, all claiming that some unholy union of different species—a lion, snake, and eagle, for instance—has produced a dragon, a sphinx, a griffin, a medusa. In Greek mythology we find satyrs, overly lustful woodland goat-men, and the perfumed and tuneful sirens, bird-women who lured men to their doom. There are Chinese tales of families that descend from the marriage of a shape-shifting dragon and a human. Siberian shamans owe their magical power to the marriage of men and swans. Native American lore declares that the first people of the Earth were part animal. In fairy tales brides and grooms marry animals. Often the chimeras (mermaids, for instance) exist at the limits of the known world, where heroes and explorers go to prove their courage. Despite the countless children's books filled with delightful thinking, talking, personality-ridden animals, the idea of a real-life part-human being trapped in the body of another animal seems diabolical to most people, so horrifying that it was the Greek gods' favorite way of punishing humans.

As exotic as it sounds, we already have a great many natural chimeras among us, including all the people who secretly harbor Neanderthal and Denisovan genes. We absorb other people all the time. When we pass along a cold sore or flu, the virus carries some of our protein and releases it inside the other person, where the immune system stows it for future reference. HIV and other retroviruses are especially good at installing pieces of one person's DNA inside another person's chromosomes. By exchanging body fluids we even swap gene fragments with our partner, and become a chimera as our self starts including bits of their immune system. We don't just get under a mate's skin, we absorb him or her. As the immunologist Gerald N. Callahan explains, we're probably swapping gene fragments with other people "a lot more often than we realize. Infection becomes communication, memorization, chimerization. Over the course of an intimate relationship, we collect a lot of pieces of someone else. Until one day what remains is truly and thoroughly a mosaic, a chimera—part man, part woman, part someone, part someone else."

However the affair turns out, we're invisibly changed for having known each other. This may not be a pleasant thought if the DNA belongs to an old flame you never want to see again as long as you live and can't bear the thought of hauling around in your car, let alone your cells. Best not to dwell on that. Think instead of Mom's DNA, or a sweetheart's, still alive inside you as a miniature portrait.

The human chimeras known as twins provide a glimpse of how confusing a world full of clones might be, but twins are too numerous to regard as oddities. At the end of her life, my mother needed regular bone marrow transplants, through which she had donor cells in her bloodstream. By that time, she was already a chimera, because moms retain cells from their fetuses. She would also have stored cells from my father, who stored her cells, too. But, to the best of my knowledge, he didn't contain cells from pig valves, cat gut, or monkey glands—though the last might have been a temptation, since it was in high vogue when he was a young man.

In the pre-Viagra 1920s, men hoping to improve their virility flocked to the French surgeon Serge Voronoff, who grafted thin slices of monkey testicles onto their scrotums. Later he transplanted monkey ovaries into women, including, allegedly, the U.S. coloratura soprano Lily Pons, who was a frequent guest at his monkey farm on the Italian Riviera. Over five hundred such operations brought Voronoff fame and great wealth, until, in time, he was denounced as nothing more than a witch or magician. But people didn't fret about hitching their testicles and ovaries to monkey glands.

Today, the monkey gland craze is as passé as Rudolph Valentino, but so many of us have pieces of other nonhuman creatures inside us, it's surprising that we don't inadvertently oink, clop, or bleat in embarrassing moments. We think nothing of strolling around with cow and horse valves in our hearts. Raising genetically modified pigs that are more compatible with human tissue, we harvest the blood-thinner heparin from their intestines, and insulin from their pancreases. The fibrous tissue in the spaces *between* the cells in a pig's bladder, once viewed as mere cushions, are so rich with growth factors that they're used to "fertilize" war-ravaged human muscles and help them regrow.

When Corporal Isaias Hernandez, a nineteen-year-old marine deployed in Iraq, had 70 percent of his thigh muscle torn off by a roadside bomb, doctors assumed they'd be amputating the leg. The remains of his thigh looked to him like a half-eaten meal at a Kentucky Fried Chicken restaurant: "You know, like when you take a bite of the drumstick down to the bone?" Quickly scarring over, his thigh sparked constant pain, and doctors prescribed amputation followed by a prosthetic as his only hope.

Then he became a chimera. Volunteering to be part of a clinical research trial, in 2004, he allowed surgeons to insert a paper-thin slice of pig's bladder, known as extracellular matrix, into the ragged thigh muscle. It began to regenerate. Today, without pain, he—like others—uses a regrown thigh to walk, sit, kneel, bike, climb, and enjoy a normal life. He'll always be part pig, the part his surgeons refer to affectionately as "pixie dust."

Is there much difference between ingesting and implanting? We swallow snake and spider venom, and gila monster spit to calm an unruly heart, and cone shell venom for pain. For birth control, millions of women ingest mare's urine. Our foremost antibiotics come from a cavalcade of fungi. Then there are the coatings, capsules, and liquid additives that go into medicines, concocted from the skin, cartilage, connective tissues, and bones of animals. If we're comfortable with implanting horse valves in our faulty hearts and pig tissue in our thighs, if we get past the basic idea of raising animals to butcher for their organs and amending our bodies with pieces from lower orders, what else might we think of? Borrowing a spare stomach from a cow so that we can digest food more quickly and lose weight?

Maybe embedding parts from other animals doesn't seem to bother us because, on the atomic level, we're living beings composed of nonliving parts. Hence the graveside reminder, "from dust to dust." Maybe we see it as the ultimate domestication of animals and taming of the soil, which we began long ago in our collective memory, little by little widening their uses. Gradually we've gone from animals sleeping under our roof to animals sleeping under our ribs without feeling alarmed. *Oh, that again, the cow is in my bone house*. Maybe in our desperate hours we gladly extend the idea of kinship from, say, my brother's kidney to a sheep's kidney.

Man-made chimeric creatures are a staple in laboratories—mice and other animals bred or grafted with human immune systems, kidneys, skin, muscle tissue—as a common way to study human diseases. Scientists have created sheep with organs that are 40 percent human, monkeys with part-human brains, and mice in which a quarter of the brain cells were human (fortunately they still behaved like mice, but who knows what strange mists galloped across their thoughts). Yet people balked when Japanese scientists announced that, given a year, they could grow a perfect human heart or kidney by tucking a human stem cell into a pig's embryo, then lodging the embryo in a healthy pig's womb. Pig valves in humans, no problem. But a pig with human organs?

It's a sign of our times that the problem with "chimeric embryos" isn't technological but ethical. It's doable, but it's not permissible. Nations would have to agree on what a human being *should be*, and that's not so obvious anymore. For the first time, we're asking ourselves: how far are we willing to engineer the world and ourselves? We still feel human when partially enhanced by prostheses, somewhat chimeric, or controlling wearable technology by eyelash-flicks or thoughts. The question has become one of degree. How replaceable are we, yet still legally and attractively human? And where is the line of disgust between enhanced human and monstrous?

Canada has passed the Assisted Human Reproduction Act, which bans the creation of chimeras. The bioethicist Françoise Baylis of Dalhousie University in Halifax, Nova Scotia, helped draft Canada's guidelines on chimeras.

"We don't treat all humans well, and we certainly don't treat animals well," she insists. "So how do we treat these new beings?"

In the United States, the National Academy of Science permits chimeras but warns against allowing chimeras to breed, because breeding two part-human chimeras could potentially lead to the grotesque (though almost certainly fatal) possibility of a human embryo growing inside another animal. Remember Rome's fabled origin, when Romulus and Remus were raised by wolves? Suppose a wolf actually gave birth to a human? Or a sheep did? Almost ten years ago, Esmail Zanjani of the University of Nevada, Reno, announced that he had injected human stem cells into sheep embryos halfway through gestation, and the lambs emerged with human cells throughout their tissues. And not just a few cells. Some of the organs were nearly *half* human. Only the organs. No two-legged sheep with opposable thumbs emerged. Staring at them in photographs, I found they looked eerily human, with long faces, jelly roll falling over the forehead, and down-turned eyes. Would dogs detect an odor both human and sheep?

What scientists still don't know is if transplanted human stem cells would change an animal's inherent behaviors, attributes, or personality. As bioethicists rightly argue, the last thing we need is the horror of humanized monkeys or other animals. With less than one-thousandth the brain volume of humans, there's little danger of mice developing our cognitive abilities. But in an animal closer to us on the evolutionary tree, say, a chimpanzee or bonobo, the merger might just work, especially if the DNA were mixed in the earliest stages of development. What would the orangutan Budi make of monkeys with part-human brains, I wonder?

A laboratory chimera poses a moral paradox. The more human its cells, the better it will serve for testing human cures. Too human, and it's trapped between worlds, a claustrophobic prisoner. Writing in 1876, when the Industrial Revolution had really begun to pick up steam, one British novelist warned of just such a possibility.

In H. G. Wells's classic novel *The Island of Dr. Moreau*, a shipwrecked man, rescued by a passing boat, relates a gruesomely fascinating tale of escaping from a nameless Indonesian island populated by sentient monsters whom Dr. Moreau has created through transfusions, transplants, grafts, and other bizarre techniques to create human-animal chimeras. They're hyena-swine, hog-men, leopard-men, apemen, little sloth people, and other "Beast Folk," some of whom have founded their own colony in the jungle, worship Moreau, and have evolved moral bylaws. The novel shocked Victorian England, which was reeling from a slew of new technologies and from Darwin's idea that humans descended from apes. The vogue for vivisection provoked controversy, as did eugenics, and the ethical limits of scientific experiments. Wells's novel brought all of those into question, and also explored British colonialism, the essence of identity, the depravity of torture, and maybe most of all the peril one faces by interfering with nature. In later years, Wells described the wildly successful novel as "an exercise in youthful blasphemy."

Gene splicing and bioengineering would not appear for a hundred years, but Wells foresaw some of the ethical dilemmas they might pose a little later in the Anthropocene. Suppose, by accident or design, a subhuman chimera emerged, something more intelligent than other animals, but less so than humans? What purpose would it be expected to serve? What sort of home would it find in our society? Would it be relegated to a lower caste? Under what circumstances should we consider a man-made chimera human? What inalienable rights would it possess?

DNA'S SECRET DOORMEN

Swinging gamely among the fire-hose vines at the Toronto Zoo, Budi isn't a cyborg or man-made chimera, and no human has reknit his DNA. He's just a frisky orangutan kid, an emissary from the wild. But we're starting to regard his physical nature (and our own) in radically new ways that connect and redefine us. Only the knowledge and what we can do with it are new. The rest is ancient as the family tree we share.

A YOUNG WOMAN with chestnut hair is seated in front of me in the cinema, slouched down, watching Stanley Kubrick's *2001: A Space Odyssey*. On the art-house screen, a vegetarian ape idly fingers the scattered bones of a fallen antelope. Slowly an idea begins to take shape. Picking up a bone, he raises it over his head and smashes it down on the rest of the skeleton, over and over, striking and shattering in an orgy of violence, while the vision of a tasty tapir flickers through his head and the pounding chords of Richard Strauss's *Thus Spake Zarathustra* drive home the message: Man the Hunter is born.

A day later, the ape man uses his weapon to kill the leader of a rival band of apes, while the Strauss soundtrack grows orgasmic with a new drama: the blow-by-blow chords of war. From there, Kubrick treats us to human evolution, artificial intelligence, alien life, and technological pageantry. Cascading into the spacefaring future, we find an astronaut vying with a sentient, mentally disturbed computer (which he subdues with a tool far subtler than an antelope bone). Reaching the apogee of his fate, he's transfigured in a process that's too advanced for us relative cavemen (in the cave of the movie theater, anyway) to distinguish from magic. As the credits roll like blankets of stars, rising houselights return us to Earth and a human saga and future that seem all the more epic.

When the chestnut-haired woman gets up to leave, one strand of hair remains on the back of her seat. From that tiny sample, someone could peruse the DNA and know if it belonged to a human female or an Irish setter or a fox, and find clues to her identity: ethnic background, eye color, likelihood of developing various diseases, even her probable life span. One would assume that she has little in common with a mouse or a roundworm, and yet they have a similar number of genes. She's intimately related to almost every creature that walks, crawls, slithers, or flies, even the ones she'd find icky. Especially those. She shares all but a drop of her genetic heritage with spineless organisms. But that drop really counts.

Thanks to the Human Genome Project's library of our roughly twenty-five thousand protein-coding genes, available via the Internet to anyone with a yen to peruse it, a micro-stalker could analyze the rungs of our redhead's DNA, creeping up its spiral ladders, and discovering all sorts of juicy nuggets. Some of the micro-portrait he finds will be quite recent, because by hogging and restyling the environment we've altered plants, animals, single-celled organisms, and ourselves. Her DNA will show a panoply of revisions, indicative of our age, which we've either stage-managed or accidentally caused. Could the pollutants we use, and the wars we wage, really change our DNA and rewire the human species?

She knows they can, because in her college curriculum, Anthropocene Studies, she's read research linking exposure to jet fuel, dioxin, the pesticides DEET and permethrin, plastics, and hydrocarbon mixtures to cancer, and not just in the person who had contact with it but for several generations. She's learned how the arsenic-polluted drinking water in the Ganges delta in Bangladesh can lead to skin cancer, as can workplace exposure to cadmium, mercury, lead, and other heavy metals. Although she was tempted to spend her junior year abroad in Beijing, she's having second thoughts now that a peer-reviewed *PLOS ONE* study ties life in smog-ridden cities to thickening of the arteries and heightened risk of heart disease. What clinches it is this headline in *Mail Online*: "China starts televising the sunrise on giant TV screens because Beijing is so clouded in smog." Below it, a video shows a scarlet sunrise on an LED billboard in Tiananmen Square, completely encased by thick gray air, as if the sun were on display in a museum. Several black silhouettes are walking past it on their way to work, some wearing masks. As a daily jogger, she'd be inhaling a lot more pollution than most people, and she figures her genes have already been restyled just by growing up among the master trailblazers of the Human Age.

But she is tempted to read the book of her genes, and discover more about her lineage and genetic biases. For a truly personal profile, all our redhead would need is a vial of her blood and between $100 and $1,000. Such companies as Navigenics or 23andMe will gladly provide a glimpse of her future, a tale still being written but legible enough for genetic fortune-telling. She may have a slightly higher than normal risk of macular degeneration, a tendency to go bald, a gene variant that's a well-known cause of blood cancer, maybe a different variant associated with Alzheimer's, the family bane. If she read the report herself, she might not handle that information well. It could kindle needless worry about ailments that will never materialize, or it might warn of an impending but treatable illness, or predict a serious, disabling disease like Huntington's. As a supposed calmative, such tests are usually marketed as a "recreational" exercise, to discover if you're part Cherokee or African or Celt, or Neanderthal, or even related to Genghis Khan, as I may well be.

My mother always said I must be part Mongolian, because of my lotus-pale complexion and squid-ink-black hair. *Something you're not telling me?* I was tempted to ask. But I knew she'd visited Mongolia with my father long after I was born. What I didn't know is that one out of every two hundred males on Earth is related to Genghis Khan.

An international team of geneticists conducting a ten-year study of men living in what once was the Mongolian empire discovered that a surprisingly large number share the identical Y chromosome, which is passed down only from father to son. One individual's Y chromosome can be found in sixteen million men "in a vast section of Asia from Manchuria near the sea of Japan to Uzbekistan and Afghanistan in Central Asia."

The likeliest candidate is Khan, a warlord who raped and pillaged one town after another, killing all the men and impregnating the women, sowing his seed from China to Eastern Europe. Though legend credits Khan with many wives and offspring, he didn't need to do all the begetting himself to ensure that his genes would flourish. His sons inherited the identical Y chromosome from him, as did their sons and their sons' sons down a long, winding Silk Road of legitimate and illegitimate progeny. His equally warlike oldest son, Tushi, had forty legitimate sons (and who knows how many misbegotten), and his grandson Kublai Khan, who figured so large in Marco Polo's life, had twenty-two.

Their genes scattered exponentially in an ever-widening fan, and the process really picked up speed in the twentieth century, when cars, trains, and airplanes began propelling genes around the planet and stretching the idea of "courting distance," which used to be only twelve miles—how far a man could ride on horseback to visit his sweetheart and return home the same day. Now it's commonplace to have children with someone from thousands of miles, even half a world, away.

Khan wasn't trying to create a world in his image; his fiercest instincts had a mind of their own, and his savage personality spurred them on. Most people don't run amok on murderous sprees, thank heavens, but history is awash with Khan-like wars and mayhem. In their wake, gene pools often change. One can only surmise that wiping out the genes of others and planting your own (what we call genocide) must come naturally to our kind, as it does to some other animals, from ants to lions.

Typically, wandering male lions attack a pride, drive off the other males, and kill their offspring. Then they mate with the females, ensuring that only the invaders' genes will flourish. A colony of ants will slaughter millions of neighbors, provided they're not family (somehow they can spot or whiff geographically distant kin they haven't met before). Human history is riddled with similar dramas, but that doesn't justify them. They were, and are, war's legacy, an unconscious motive, not a blueprint for action.

Except once. During World War II, Hitler and his henchmen devised an agenda, both political and genetic, that was nothing less than the Nazification of nature. The human cost is well known: the extermination of millions while, in baby farms scattered around Europe, robust SS men and blond, blue-eyed women produced thousands of babies to use as seed stock for Hitler's new master race. What's little known is that their scheme for redesigning nature didn't stop with people. The best soldiers needed to eat the best food, which Nazi biology argued could grow only from the purest of seeds. So, using eugenics, a method of breeding to emphasize specific traits, the Nazis hoped to invade the genetic spirals of evolution, seize control, and replace "unfit" foreign crops and livestock with genetically pure, so-called Aryan ones.

To that end, they created an SS commando unit for botanical collection, which was ordered to raid the world's botanical gardens and institutes and steal the best specimens. Starting with Poland, they planned on using slave labor to drain about a hundred thousand square miles of wetlands so that they could farm it with Aryan crops.

Draining the marshes might well lower the water table and create a dust bowl, and it would certainly kill the habitat of wolves, geese, wild boar, and many other native species, but despoilers rarely see downstream from events.

Elsewhere, the Nazis proposed planting forests of oak, birch, beech, yew, and pine to sweeten the climate so that it was more favorable for their own oats and wheat, and they spoke openly about reshaping the landscape to better suit Nazi ideals. That revision included people, railways, animals, and land alike, even the geometry of farm fields (no acute angles below 70°), and the alignment of trees and shrubs (only on north-south or east-west axes). Today, though we deplore genocide it stubbornly persists, and we may have our work cut out for us because it seems to tap a deeply rooted drive. It's bone-chilling how close the Nazis came to a feat of genetic domination that dwarfs all of Genghis Khan's exploits.

The human DNA that Olivine finds in future days will show some lineages, like Khan's, triumphing through war, and others succeeding because of geography, religion, politics, fashionable ideals of beauty, and elements native to our age such as giant factories and workplaces, cars, jet travel, Internet and social media, the jammed crossroads of megacities, and widely available birth control and infertility treatments.

When my mother teased about my being part Mongolian, she may have been right, since Genghis Khan and his clan reached into Russia. But I like knowing that the farther back one traces any lineage the narrower the path grows, to the haunt of just a few shaggy ancestors, with luck on their side, little gizmos in their cells, and a future storied with impulses and choices that will ultimately define them.

The noble goal of the Human Genome Project is to use such knowledge to find new ways to understand, treat, and cure illness. In that sense, it's a group portrait of us as a species, realized at last, a mere fifty years after Crick, Watson, and Franklin decoded the double-helix design of DNA. The only thing more unlikely than

DNA itself, nature's blueprint for building a human being, is our ability to decode it. Thus far, it's our greatest voyage of discovery, and we're still scouting its spiral coves.

IN NORRBOTTEN, THE northernmost province of Sweden, the reindeer outnumber humans, and shimmery green veils of northern lights spiral up from the horizon like enchanted scarves. In summer, crops ripen under a ceramic sun; moon-shadows haunt the ice-marbled winter nights. Although the citizens can travel by car today, in the past they relied on foot or horse power to carry them to grace at an early-fifteenth-century church in the ancient settlement of Gammelstad, where they eased their isolation and replenished hope.

Getting there was only half the pilgrimage. Needing rest before the formidable trek home, each family retired to their tiny wooden house near the church, painted red with windows and doors picked out in white. Some bore grass roofs. Delicate lace curtains hemmed the frost-curled windows, and stout shutters sealed out the warring tempests. Doors were adorned with a pyramid motif, a legacy from pagan antiquity admired for its stark symmetry, and reinterpreted as a Christian altar lit by sacrificial fire.

In such a remote frontier, the human population thinned to only about six people per square mile, and farmers crafted what they needed, from harnesses to nails. Neighbors helped neighbors, and married neighbors. But if the harvest failed—as it did with alarming frequency—rescue lay too far away. Railways didn't venture that far north, even at the height of the Industrial Age when iron horses snorted soot across many frontiers, and in any case locals spoke a dialect unintelligible to other Swedes.

Gammelstad's church, plus the rows of red bungalows clustered behind it, are part of a World Heritage Site that also includes the remains of a six-thousand-year-old Stone Age settlement in the heart of town. Tourists are today's pilgrims, closely followed by geneticists. It's an unlikely setting to be at the center of a revolution in medicine, and yet it holds an important key to the health and longevity of everyone on Earth.

In the nineteenth century, Norrbotten's fickle climate bred many lopsided years of surfeit or famine, with no way to foretell the fate of the crops. People either nearly starved or died. For example, 1800, 1812, 1821, 1836, and 1856 all were years of deprivation, when crops totally failed (including staples like potatoes and grains for porridge), farm animals died, families suffered pounding hunger and malnutrition, and underweight babies entered a lean world with even leaner prospects. But in 1801, 1822, 1828, 1844, and 1863, on the other hand, the weather sweetened and food leapt from the soil in such abundance that families thrived, the economy bloomed, and for many people overeating became a pastime.

If we jump to the 1980s and step across the North Sea to London, we find the prestigious medical journal *The Lancet* publishing studies that highlighted the importance of womb-time, linking a mother's poor diet during pregnancy with her child's higher risk of heart disease, diabetes, obesity, and related illnesses. This was a revelation to the medical community and a warning to parents.

According to Darwin's theory of natural selection, a child is born with a genetic blueprint that has evolved over millennia. All the hard-luck times its parents faced might be taught as life lessons, but they aren't hereditary; they won't alter a child's chemistry. Thinking otherwise was a delusion mocked and dismissed in the eighteenth century, when the naturalist Jean-Baptiste Lamarck (the man who coined the word "biology") posed a theory that parents could pass along acquired traits to their offspring. In his most famous example, a giraffe reaches achingly high into the treetops each day to feed on tasty leaves, which ultimately lengthens its neck, and then its offspring inherit longer necks and stretch them even more by mimicking the parent's behavior. Smart, keen-eyed, and right about many aspects of botany and zoology, including the dangerous idea that new species arise naturally through evolution, Lamarck was wrongheaded about giraffe necks and heritable traits. According to his logic, if a blacksmith grew anvil-hard arms from a lifetime of heavy hammering, his offspring would inherit equally burly muscles. It's fun to think what such a world would look like—a mismatched crowd of animals within each species and the enviable ability to *will* traits to one's offspring. Practice wouldn't be needed—you could inherit your pianist dad's spiderlike dexterity with his fingers or your bicycling mom's loaflike quadriceps.

Darwinian evolution teaches us that genetic changes in DNA unfold with granular slowness over millennia; no individual can erase or rewrite them in his lifetime. Genetic mutations come and go, and if one is harmful or useless for survival, it tends not to linger. But if it's beneficial it equips the animal with an edge, a better chance at surviving long enough to breed, and then the mutation empowers the animal's offspring and their offspring in turn, passing the winning trait along to future generations in quite a sloppy way, all things considered. In time this fluky mechanism leaves the world with only those animals best suited to their different habitats.

That's the accepted theory, proven in countless experiments, and there's no reason to doubt it. But what if that isn't the whole story? Eclipsed by Darwin, Lamarck seems to have been right at least in spirit, a reality that has stunned much of the scientific world. What makes a paradigm shift so shifty is that you don't see it coming. Then it suddenly pulls a mental ripcord and your mind plummets at speed. A new paradigm blossoms overhead, the freefall slows to float, and the world becomes visible once again, but from a new perspective. "Creative insight," we call this parachute flare with discovery.

After Lars Olov Bygren, of the Karolinska Institute in Stockholm, read the *Lancet* articles, he began to wonder about the nineteenth-century children of Norrbotten who had alternately starved and binged. The people of that region seemed ideally isolated for a genetics study. Certainly the children would have been influenced by their mothers' nutrition *during* pregnancy, but what about all the earlier feasts and famines their parents endured—could those blemish the children's health? This was a daring question to ask, let alone pursue, since it flew in the face of Darwinian fashion. But it nagged at him until he finally decided to focus on ninety-nine children born in Överkalix in 1905, relying on a wealth of historical data. Why choose those mountain bluffs and chanting shores?

"I grew up in a small forested area, ten miles north of the Arctic Circle," Bygren explains.

A slender man with gray hair and round glasses, he walks thoughtfully among the headstones in the church cemetery, where tall stalks of sunlit grass surround some of the young who died for lack of grain a hundred years before. The headstones bear such familiar names as Larssen and Persson, the English equivalent of Smith and Jones; Bygren probably played with some of their relatives as a boy. Bluebells and daisies flower naturally between the stones, and some graves are adorned with gaudier store-bought flowers.

"We have an expression," he says with a laugh. "Dig where you stand!"

After a moment, Bygren adds, "We are really data rich." Data rich even when crop poor. "Everything happening in the family was recorded."

Ever since the sixteenth century, the clergy has kept a fastidious ledger of births and deaths (with causes), as well as land ownership, crop prices, and harvests. Thanks to the clergy's meticulous record-keeping, Bygren was able to gauge how much food was on hand when parents and grandparents were children. Överkalix provided a natural experiment where he could follow isolated families as they tumbled forward in time.

Common sense hints that if you're creased by trauma, are a junk food junkie, or spend your days staring at lit screens while spring offers the likes of pink-petaled magnolias ruffling their flamingo feathers in the breeze, it may make you miserable and unhealthy, but it won't affect the DNA of your unborn children. They may inherit your curly hair, gray eyes, musical ear reliable as a tuning fork, thin porcelain skin, or risk of a genetic disease such as Huntington's, but they won't suffer from your accidents and misdeeds. Their DNA won't be damaged by your rotten choices in lifestyle, nor can you pass on all the wonderful feats you've accomplished, the wonders your senses have soaked in, the perils you've avoided. In that sense, they're born with a clean slate and will become entangled in their own dramas and make their own questionable choices. Certainly they won't be obese, get diabetes, or die young just because a grandparent binged during one tantalizingly rich harvest season after a year of brutal gourd-bellied hunger. Evolution doesn't work that way or that fast—end of story. Or is it?

What Bygren found was quite different. He and the geneticist Marcus Pembrey, of University College London, began collaborating on groundbreaking studies that raised eyebrows and led to such headlines as "You Can Traumatize Your DNA," "You Are What Your Grandparents Ate," "Nurture Matters," and "The Sins of Our Fathers" (in Exodus, God speaks of "visiting the iniquity of the fathers on the children and the children's children").

The ultimate immigrants, babies arrive in this world from a far country with no dry land, lugging helical clouds of ancient DNA, primed for survival, but seemingly ill equipped to face sudden changes in the environment. Yet it is possible to warn children and grandchildren about recent dangers. Episodes of near-starvation—or other extreme changes in the environment—tag the DNA in children's nascent eggs and sperm. Then years later, when they have children of their own, new traits emerge, not because the traits serve the species well, but because of the parents' specific stresses long before their children were conceived.

When Bygren looked at the children of Överkalix, he was surprised to discover that boys between the ages of nine and twelve who gorged during a bountiful season, inviting diabetes and heart problems, produced sons and grandsons with shorter life spans. And not by a negligible amount. Both sons and grandsons lived an average of thirty-two years less! In contrast, the boys who suffered a hunger winter, if they survived and grew up to have sons of their own, raised boys with health benefits—four times less diabetes and heart disease than their peers and life spans that averaged thirty-two years longer. Later studies found similar results among the girls, though at a younger age, since girls are born with a bevy of eggs and boys develop sperm in the prelude to puberty. During these growth windows, ripening eggs and sperm seem to be especially vulnerable to intel about the environment. Like a computer's binary code, the marks tell switches to turn on or off in the cells. Then eggs and/or sperm ferry the message to the next generation, where they may indeed be lifesaving. On the other hand, they could usher in the onset of disease by equipping someone for a world that no longer exists. Problems detonate when one is biologically prepared for a radically different environment.

"The results are there," Bygren says, solid as the iron ore enriching the folds of Norrbotten. "The mechanisms are not so clear."

Shocking though the idea was, the evidence plainly showed that it only took a single generation to make indelible changes. That year of gluttony as a child set in motion a biological avalanche in the cells, dooming the children's as yet unimagined grandchildren to a host of illnesses and vastly shorter lives than their peers. It's as if they had inherited a genetic scar.

How and why this evolutionary sidestep happens is the focus of epigenetics, a new science that puts all the old-fashioned college debates about nature or nurture on the Anthropocene scrap heap of outmoded ideas. It also lays a heavier burden on the shoulders of would-be parents. Apparently, it's never too soon to begin worrying about the health of your grandchildren.

The implications are staggering. Up until now, inheritance was a tale told by DNA; it lay exclusively in the genes. In the watch-how-you-step, deep-nurture world of epigenetics, proteins tag DNA by coiling around it, pythonlike, squeezing some genes tighter and loosening others, in the process switching them on or off, or leaving them on but turning the volume up or dialing it down to a whisper.

Changes in our genome took millions of years, but the epigenome can be changed quickly, for example, by simply adding a tiny methyl group (three hydrogen atoms glued to one carbon atom) or an acetyl group (two carbons, three hydrogens, and an oxygen). This "methylation" turns a gene off, and "acetylation" turns a gene on. Environmental stresses flip the switch, which makes sense, since in theory it prepares offspring for the environment they're going to find. Diet, stress, prenatal nutrition, and neglect create especially strong marks, whose influence can be either good or bad. How the marks fiddle with your genes may be deadly in the long run, or may prolong your life. Exercise and good nutrition leave beneficial tags, smoking and high stress pernicious ones.

What changes isn't the tool but how it's used. It's like the difference between wielding a hammer to tap in a picture nail or to smash a hole in a wall. Nature is thrifty, recycling the basics. It's as if DNA were a tonal language using the same consonants and vowels, but speaking them with different inflections. In Mandarin Chinese, the world's most widely spoken tonal language, how you voice the word "man" determines whether you mean "slow" or "deceive." Exactly the same DNA funds heart, pancreas, and brain cells, yet they finesse different tasks. As genes are switched on and off, made to shout or whisper, their meaning and purpose shift. That's why it's merely an embarrassment that we have fewer genes than plants and nearly the same genes as chimpanzees. Gifted with the same libretto of genes, life forms intone them differently, and our own cells morph into skin, bone, lips, liver, blood. Epigenetics is providing clues to how this tonal magic is performed.

Pembrey's fascinating hypothesis is that the Industrial Age ushered in a flood of rapid-fire environmental and social changes, and while genetic evolution struggled to keep up with them, it couldn't adapt that fast. The speed of change was unprecedented, and our genes don't evolve in just a few generations. But certain "epigenetic tags" clinging to those genes could. So the pesticides or hydrocarbons your great-grandmother was exposed to when she was pregnant may heighten the risk of ovarian disease in you, and you in turn might pass that risk on to your grandchildren. Ovarian cancer has been increasing to affect about one in seventy-three women over the past few decades, and environmental epigenetics offers a plausible reason why.

We only exist in relation to others and the world. This dialogue, a three-ring circus among the genes, a perpetual biological tango performed by multitudes, deserves a better name than the unwieldy crunch of "epigenetics," but the word is springing from many more lips as doctors search for clues in both a patient's environmental exposure history and that of his parents.

"We're in the midst of probably the biggest revolution in biology," says Mark Mehler, chair of the Department of Neurology at Albert Einstein College of Medicine. "It's forever going to transform the way we understand genetics, environment, the way the two interact, what causes disease. It's another level of biology, which for the first time really is up to the task of explaining the biological complexity of life."

"The Human Genome Project was supposed to usher in a new era of personalized medicine," Mehler told the American Academy of Neurology at its annual meeting in 2011. "Instead, it alerted us to the presence of a second, more sophisticated genome that needed to be studied."

Despite the DNA of twins, for example, they're never perfect matches. If one has schizophrenia, the odds of her twin developing it are only 50 percent, not 100 percent as one might assume since they have identical genes. Twins have become an important part of epigenetic studies. So have children of Holocaust survivors, Romanian orphans who weren't held and comforted enough, and children with stress-rattled or neglectful caregivers. From psychiatric epigenetics we're learning how important a mother's mood is to the fate of her fetus. The chemicals that swaddle and seep through a fetus can influence its future health, mood, and life span.

In 2004, Michael Meaney, whose lab at McGill University was studying maternal behavior, published his findings in *Nature Neuroscience*. Good mother rats, who licked their fourteen to twenty pups often and with care during the first week of life, produced nice calm pups. Standoffish mother rats who didn't lick their pups much or neglected them entirely produced noticeably anxious pups. And as adults, the next generation of female rats mirrored their mothers' behavior.

"For us, the Holy Grail was to identify the path that was being altered by this licking behavior," Meaney says. "We identified one small region on the gene that responds to maternal care and directs changes in the brain cells."

Meaney's work is now looking at human child development. A highly stressed pregnant mother floods her fetus with glucocorticoids, which can reduce birth weight, shrink the size of the hippocampus (memory's estate), and cripple the ability to deal with stress. Yet, as Meaney is the first to point out, many underweight babies turn out fine, which suggests that postnatal care must be able to reverse the ill effects. Environment and nurture *do* matter, and it doesn't take long for their influence to show. In colonies of the desert locust, individuals are naturally shy and nocturnal. But if the population swells and overcrowding occurs, the densely packed grasshoppers give birth to gregarious, diurnal young. Or, in bird studies, if the mother lives in "socially demanding conditions," holding high social rank, for example, her androgen level climbs. That leads to increased androgen in her eggs and produces more competitive chicks.

Is our moviegoer among the poor? As the health consequences of poverty drift into social medicine's sights, ongoing human and animal studies link either enriched environments or impoverished ones to the health of children and grandchildren. Studies of the Dutch Hunger Winter in 1944–45 reveal that prenatal hunger can lead to schizophrenia and depression. In a U.K. study, poor prenatal nutrition is tied to a trio of risks for heart disease in older adults.

In other research, mothers who lived through the stresses of hurricanes and tropical storms while they were pregnant were more likely to have autistic children. Even if as an adult the redhead makes all the healthy choices, is happy during her own pregnancy, and becomes a doting mom, her child can still suffer from the stresses its neglected grandmother endured during the Great Depression. Or its grandfather suffered in Vietnam as a young recruit before she was even conceived. What her grandparents ate for breakfast matters.

Did our redhead's father take Viagra? Thanks to such drugs, much-older men are siring offspring. What effect could so many older fathers, and aging genes, have on us and future generations? One unexpected finding is that, for some reason, older fathers endow their offspring with longer telomeres (which cap the ends of chromosomes like the tabs at the end of shoelaces do to keep them from fraying), a part of the gene that controls life span. So the children may live longer. On the other hand, older fathers are blamed for passing on mutations that can lead to such dreaded disorders as autism or schizophrenia. Dad's diet was also important; if he was gluttonous, she may be more likely to develop diabetes. Fortunately, our moviegoer inherited telomeres long as a summer night.

None of this happens by unzipping and altering the codebook of DNA, yet it's inherited by offspring. Epigenetics is the second pair of pants in the genetic suit, another weave of heredity, and although revising someone's genome is hard, it's relatively easy to change an epigenome. The marks are profound but not permanent. As a result, the field holds limitless promise.

"Genes can't function independently of their environment," Meaney says. "So every aspect of our lives is a constant function of the dialogue between environmental signals and the genome. The bottom line seems to be that parental care can have an even bigger impact than we ever dreamed on our children's lives. We're just starting to learn what that means."

Yes, a trauma in your mother's childhood could affect your health, and the health of your child—but it's also reversible, even as an adult. In the McGill study, researchers were able to undo the chilly behavior of the second generation of mother rats by using epigenetic drugs to turn genes on or off.

The great promise of epigenetics is the possibility of curing cancer, bipolar disorder, schizophrenia, Alzheimer's, diabetes, and autism by simply flipping the switches that tell some genes to wake up or work overtime, and others to lighten up or nap. Can we really hypnotize our genes like that, canceling out bad behavior and sparing innocent offspring before we plan to have any? The consensus is yes. Scientists have begun developing drugs such as azacitidine (given to patients with certain blood disorders) capable of silencing bum genes and spurring on healing ones. Many illnesses, such as ALS and autism, appear to be epigenetic, which puts them within reach. Three different types of epigenetic drug therapy are being actively investigated for schizophrenia, bipolar disorder, and other major psychoses. The FDA has already approved several epigenetic drugs, and in 2008 the National Institutes of Health (NIH) declared epigenetics "central" to biology and committed $190 million to understanding "how and when epigenetic processes control genes." The Human Genome Project, completed in 2003, rightly celebrated as a wonder of human ingenuity, had only twenty-five thousand genes to map. The epigenome is much more complicated, with millions of telltale marks. So a full epigenome will take a while, but an international Human Epigenome Project is under way.

The good news is that these are problems with possible, if not simple, solutions: ban more environmental toxins known to trigger epigenetic havoc; work harder to ease famine, reduce poverty, and repair the ravages of war; and help people understand the long-term impact of their actions and the vital role that nurture plays in their families, societies, and environment. Genes may remember how they once behaved in parent and grandparent cells, but, fortunately, they can also learn healthy behaviors, based on use, just as muscles do. What you experience in your lifetime will become a vital part of your child's legacy. Your adult experiences can rewire your genes in positive ways, and just as startlingly, the nurturing you do for friends, sweethearts, and other people's children can have lasting epigenetic effects. Once that idea registers, it changes the relationship between generations, which suddenly have everything in common, and the tapestry of the human condition grows a little more visible, thread by thread. At the level of DNA's phantom doormen, we can be connected to anyone and everyone.

There's also a moral, social, and political lesson: while humanitarian programs may seem nonessential, an extravagance of resources and spirit we can't afford, epigenetics teaches us that, on the contrary, poor education, violence, hunger, and poverty leave scars on one generation after another in a way that ultimately affects the future health and well-being of whole societies. What happens to war-torn soldiers and civilians during and after battle leaves epigenetic traces to wound future generations, adding to a country's problems, even in peacetime. The same is true of natural disasters, and we've seen plenty of both of late. Who knows what epigenetic aftermath will result? Genetic engineering may seem like a diabolical threat to us as a species, and we do need scrupulous oversight and control of such life forms. But the political and environmental choices we make—those with epigenetic repercussions—are equally powerful engines of change, ones we can often identify and fine-tune.

MEET MY MAKER,
THE MAD MOLECULE

Returning to our mystery redhead in the movie theater—what else could we learn about her from a strand of hair or blood sample? Her DNA profile, resembling a supermarket barcode, is a monumental accomplishment, but it's only a fraction of her story. For a fuller picture of her health and heredity, we would need to include the teeming seashores of her microbes, the rest of her being—in fact, more than her being. Another self, a shadow self. At any moment, she is inseparable from trillions of her single-celled, single-minded, naked companions, some of whom don't have her best interests at heart.

When she weighed herself earlier today, she may have deducted a pound or two for clothes and shoes. But did she take into account the roughly three pounds of microorganisms that inhabit her crevices and innards? Probably not. She'd need an atomic scale to start with, and anyway microbes are shifty, jumping off her peninsulalike feet and climbing aboard elbows pell-mell; they're not easy to tally.

Microbes are the most fruitful life form on Earth, colonizing all sorts of ardent unspoken strangers, creating small sulfurous rumblings in the animal belly, reveling in the smell of fish and old shoes, and leaving aftertastes in the mouth stale as bus-station sandwiches. They're also real workhorses, fluting the air until it's breathable, promoting photosynthesis on the land and in the oceans, decomposing dead organisms and recycling their nutrients. In industry, we breed them to ferment dairy products and to process paper, drugs, fuel, vaccines, cloth, tea, natural gas, precious metals; they help mop up our oil spills. We yoke them like oxen and set them to work. But just as most of the mass in the universe (94 percent) is "dark matter," this largest biomass on our planet escapes the naked eye, yet is the invisible Riviera of the visible world.

How remarkable it is that we're not only renaming our age, we're on the threshold of redefining ourselves as a completely different kind of animal than we ever imagined. For years, we thought DNA told the whole story. Instead we find that each person is a biological extravaganza of ten trillion microbes and one trillion human cells. It's amazing we don't slosh or disintegrate as we walk. Here's the thing: on a microscopic level we do, while constantly adding new microbes from other people, plumes of dust, and the plants and animals we encounter.

In only the past ten years, our picture of a human being has evolved from a lone animal to a team of millions of life forms working in unison for mutual benefit. Unrelated people may be widespread from Tierra del Fuego to Quaanaaq, but there's a movement afoot to classify human beings as "eusocial," a single unit of highly sociable life forms who can't survive all by themselves. Earth favors similar collectives—ants, bees, termites, coral, slime mold, naked mole rats, etc.—in which individuals pool their know-how to act for "the sake of the hive." Thanks to the Web and social media, we're discovering what a bustling rialto each person really is, and also how connected we all remain. Worlds within worlds, each of us is a unique ambulatory superorganism who belongs to one miscellaneous species living on the body of a colossal superorganism of a planet in a waltz of innumerable galaxies sprinkled with other Gaia-like planets and likely their own life forms percolating with untold hangers-on.

A marvel of the Human Age is that, in the past decade alone, we've mapped both the DNA in our cells and the DNA in our microbes. In the hunt we've discovered that a true view of ourselves as a life form is more untidy than we thought, and unglimpsed by most of us, a cloud of entwined bugs and human cells in a semipermeable frame. Joshua Lederberg, the Nobel-laureate biologist who, in 2000, coined the term "microbiome," defines it as "the menagerie of the body's attendant microbes." Amid the hoopla surrounding the Human Genome Project, he urged, "We must study the microbes that we carry within us and on our surfaces as part of a shared embodiment."

If the Human Genome Project was a landmark feat of discovery, the Human Microbiome Project is gene cartography's finest hour. NIH director Francis Collins compares it to "fifteenth-century explorers describing the outline of a new continent," a triumph that would "accelerate infectious disease research in a way previously impossible."

For five years, a consortium of eighty universities and scientific labs sampled, analyzed, and audited over ten thousand species that share our human ecosystem, thus mapping our "microbiome," the normal microbial makeup of healthy adults. And the quest continues.

The researchers have found that each of us contains a hundred trillion microbial cells—ten times more than our human cells. When they peered deeper and compared the genes, they realized that we carry about three million genes from bacteria—360 times more than our own human code. Among the hundred or so large groups of bacteria, only four specialize in the human body. They've been sidekicks for so long that over time our fate has fused with theirs.

So, odd as it sounds, most of the genes responsible for human survival don't descend from the lucky fumblings of sperm and egg, don't come from human cells at all. They belong to our fellow travelers, the bacteria, viruses, protozoans, fungi, and other lowlife that dine, scheme, swarm, procreate, and war all over us, inside and out. Vastly more bacteria than anything else. All alone our moviegoer could be arrested for unlawful assembly. She doesn't propel a solid body but a walking ecosystem.

They also learned that we all carry pathogens, microorganisms known to spark disease. But in healthy people, the pathogens don't attack; they simply coexist with their host and the rest of the circus tumbling and roaring inside the body. The next mystery to crack is what causes some to turn deadly, which will revamp our ideas about microbes and malady.

We've known about bacteria for 350 years, ever since a seventeenth-century Dutch scientist, Antonie van Leeuwenhoek, slipped some of his saliva under a homemade microscope, which he had crafted with lenses made from whiskers of glass, and espied single-celled organisms crawling, sprawling, flailing about in the suburbs of our gums. He named them animalcules and peered at them through a vast array of lenses (an avid microscoper, he made over five hundred).

In the nineteenth century Louis Pasteur proposed that healthy microbes might be vital, and their absence spur illness. By the time tiny viruses were discovered, only a hundred years ago, people were already driving cars and flying airplanes. But we didn't have the tools to study the every-colored, shifting, scented shoal of microbes we swim in, play in, breathe in all the day long. Some cross the oceans on dust plumes. Acting as condensation nuclei, they jostle rain or snow until it falls from clouds. Far from being empty, the air, like the soil, throbs with flecks and dabs of life, more like an aerial ecosystem than a conveyor belt for clouds.

We need to reimagine the air, not as a desolate ether but as a lively, largely invisible, ecosystem. As we peer through its glassy expanse to a far trail or up at a billowing cloud, nothing blocks our view, the whole corridor looks vacant, and yet it's a community pulsing with life. Our eyes merely slide over its tiniest tenants. The sky is really another kind of ocean, and even though we sometimes used to refer to "oceans of air," we imagined barren currents; we didn't realize how life-soaked the waves really are.

When David Smith and his colleagues at the University of Washington sampled two large dust plumes that had sailed across the Pacific from Asia to Oregon, they were surprised to find thousands of different species of microbes in the plumes, plus other aerosols, dust particles, and pollutants. All suspended and wafting around the planet, tromboning and floating, interacting with life.

In this panoramic new portrait, the Anthropocene body is no longer an entity that's separate from the environment, like a balloon we pilot through the world, avoiding obstacles, but an organism that's in constant conversation with its environment, a life-and-death dialogue on such a minute level that we're not aware of it. It recognizes the mad microscopic mosaics we really are, molecular bits who trace their origins to simple one-celled blobs, then cellular flotillas that grew by engulfing others in life's oceanic swap meet. Evolving this way and that, nabbing traits, shedding traits, we went haywire in slow motion over millions of years. Maybe our cells, however much they evolve, retain a phantom sense of those early days as colonial bodies with a shared purpose, more like amoebas or slime mold than mammals. We're beginning to accept the idea of gypsy organisms that fanfare around us, making catlike raids on each other in dark simmering thickets, species as different from one another as animals adapted to rainforest, arctic, ocean, prairie, or desert. For we, too, have hillocks and estuaries, bogs and chilly outposts, sewers and pulsing rivers for them to quarrel and carouse in.

Even inside our own cells, we house more twitchy bacteria than anything else, because our mitochondria and chloroplasts were once primitive bacterial cells. They've sponged off us for so long that they can no longer exist on their own. Some our body welcomes with open pores because they handle metabolic melodies we couldn't even hum on our own. It amazes me that we've survived with such grace, since we're born dottily deficient, lacking vital survival skills such as how to digest the very foods we eagerly wolf down. An omnivorous diet helped us endure icy forests and bright broiling terra-cotta land-scapes, but we don't have all the enzymes we need to absorb those foods; our microbes assist.

In the distant past, as Earth bloomed with primitive life, strings and mounds of twinkling single-celled bacteria discovered the mutual benefits of teamwork and became allies. Others took a bolder and more violent step—they gobbled each other up. It's only at that stage that lilacs, marine iguanas, wombats, and humans became possible. As multicelled organisms grew more and more complex, the imprisoned bacteria adapted and thrived, until they became vital cogs of each complex cell.

The consensus now among evolutionary biologists is that we can't separate "our" body from those of our resident microbes, which have been fiddling in subtle ways with our nature as a spe-cies for millions of years, and influence our health and happiness to a previously unimagined degree. Study after study is showing that microbes profoundly affect our moods, life spans, personalities, and offspring. They influence not only *how* we are but *who* we are. How strange that we feel whole, one person whom we can wash and dress and conduct internal monologues with, though most of us is not only invisible but not even what we're used to defining as human. Planet Human offers a dizzying array of habitats for the unseen and the unforeseen, the hominid and microbial.

Only very recently has the scientific community acknowledged the extent to which our microbes might indeed affect our evolution, and by *our* I mean the whole *mespucha*, as they say in Yiddish (the term in biology-speak is "holobiont"). Not just individuals but all their microscopic relatives with their relative points of view. Some hijack our free will, divert our behavior, and become matchmakers. A wasp study is offering fresh insights. By definition, members of a species can mate and produce live offspring. But researchers study-ing several species of jewel wasp (loaded with ninety-six different kinds of gut bacteria) have discovered that microbes can determine whether unions between different wasp species will succeed. When two distinct species of wasps mated, their offspring kept dying. Until recently, we would have said such a fertility problem was genetic. We know now that it can be microbial. When researchers changed the wasps' microbes, the species bred favorably and hybridized. Evolution can be detoured by a mob of hidden persuaders.

Once again from the insect world, recent experiments with fruit flies are showing another way microbes can be at the helm, and the too-real possibility that bacteria have played a vital, even scary, role in our evolution. Consider how microbes control the love life of concupiscent humans and lusty fruit flies alike. Ilana Zilber-Rosenberg and her colleagues at Tel Aviv University's Department of Molecular Microbiology and Biotechnology have discovered that the bacteria inside the gut of a fruit fly sway its choice of mates.

Fruit flies raised on either molasses or starch prefer to mate with others on the same diet. But when the flies are dosed with antibiotics, which kills the microbes in their gut, they're no longer picky and will mate with any willing male. Among fruit flies, sexy males know all the right dance moves, but they also have to smell sexy, and their pheromone-cologne is modified by the microbes inhabiting the fly. For both humans and fruit flies, the love-wizards of smell are the symbiotic microbes that brew pheromones for us, their larger hosts. Scent rules in human courtship, too, especially among females looking for a mate. Although men seldom report such fussy responses to their partner's natural smell, women so often do that it's become a romantic cliché: "There just wasn't any chemistry."

Tinker with microbes and you alter stud capital, which in turn alters the genes of the female's offspring, and so on as generations disrobe or unfold their wings. The object of natural selection isn't a single plant or animal, Zilber-Rosenberg proposes, but its whole milieu, the host organism plus its microbial communities, including all the parasites, bacteria, fungi, viruses, and other bugs that call it home.

Fruit flies make appealing test subjects because we share such a bevy of mating behaviors. The dinner date, for instance. What's the quickest way to a man's heart? Forget Cupid's arrow. According to Mom-wisdom, it's coaxed by a cozy meal, in a penumbra of pleasure that mingles the fragrant food with the cook. If men are anything like fruit flies—and who's to say they're not at times; heaven knows women are—Mom was right. For female fruit flies, a dinner date is the ultimate rush. And rush it literally is, since they only live about twenty-five days and can't afford to be shy. *Live fast and die* is their mantra, and they need a handy food supply if their large new brood is to survive. Female fruit flies prefer males who favor the same chow. Still, the males need to be in the right mood, and the females are surprisingly picky and manipulative given their short career.

During fruit fly courtship, if the microbe-milled incense is right, the male extends one mandolinlike wing and serenades the female, then engages in that style of oral foreplay many humans do, before mounting her and copulating for twenty minutes or so.

We respond to the same sweet, honeylike aromas that make fruit flies amorous, and so chemists include them in perfumes. Like an insect rubbing its wings together to croon a mating call, many a medieval troubadour used a mandolin to serenade his lady, with whom he'd dine and mate. And remember that sexy tavern scene in *Tom Jones*, in which the hero and a buxom wench devour a none-too-fresh carcass with carnal abandon? Intriguingly, if a female fruit fly spies a lone mutant (or rather a mutant mutant, say, the one *normal* fruit fly with quiet brown eyes, which would be the odd-fly-out if all the rest were bauble-eyed), the female hankers for the nonconformist. In the trade, it's known as "the rare male advantage."

For fruit flies, too, beauty is in the eye of the beholder, with their microbes adjusting the focus. Did I mention that some fruit flies have come-hither eyes? I don't mean the dozens of mosaic facets, so evocative of hippie sunglasses, but the zingy psychedelic eye colors lab folk like to endow them with, the better to study mutant genes. As a Cornell grad student, I often stopped by the fetid biology lab to admire the eggplant-blackness of the bellies, the spiky hairs, the gaudy prisms of the eyes—some apricot, some teal, some brick red, some yellow, some the blue of ships on Delft pottery. I still recall the tiny haunting eyes of the fruit flies, like the captive souls of past lab assistants, and the swooping melody of their Latin name: *Drosophila melanogaster*, which translates poetically as "dark-bellied dew sipper." Because fruit flies thrive in sultry weather (82°F), the lab offered students a warm den during those numbing upstate winters when ice clotted in beards and mittens, coeds exhaled stark white clouds, and the walkways looked like a toboggan run.

A favorite of biologists hoping to peer into the dark corners of human nature, fruit flies have it all—they're prowling for mates eight to twelve hours after birth, easy to raise, and able to lay a hundred eggs a day. Plus they share about 70 percent of human disease genes, especially those linked to neurodegenerative disorders such as Parkinson's and Alzheimer's.

However, in a sly twist, the last male the female fruit fly has sex with will sire most of her many offspring, and she chooses him only after lots of romps in the orchard or lab, based on his gift for courtship and his scent. As with most animals, from squirrels to spiders, the males pursue but the females choose, and even the lowly fruit fly can be choosy.

So is the human dinner date really just courtship feeding after all, a custom (and microbial picnic) we share with fruit flies, robins, and chimpanzees, which in our chauvinistic, I'm-not-really-an-animal way we've coyly disguised? Yes. But what's the harm in that? There's a similar meal plan among the annual hordes of Japanese beetles that tat rose leaves into doilies and shovel deep into ready-to-open buds every summer. Gardeners often spy the iridescent scarabs, in twos or crews, perched atop favorite flowers, dining and mating simultaneously. Of course, the ancient Greeks and Romans, who coined the word "orgy" and found that dining lying down leveled the playing field, enjoyed blending sensory delights with equal gusto—banquets of music, food, conversation, alcohol, and sex. As the sage once put it: "Birds do it, bees do it, even educated fleas do it." No harm at all, unless the process makes you impulsive, deranged, and deadly, which in some cases, depending on the shared microbes, sex can.

Another such culprit is a momentous if commonplace human hanger-on that also bedevils rats, cats, and other mammals and has recently been studied in harrowing detail. Spread along the edges of nature, on the boundary where humans and wild animals mix, the world population of *Toxoplasma gondii*, a particularly mischievous parasite, is ballooning with our own numbers. One way to catch the infection is to eat undercooked kangaroo meat. Kangaroo was recently approved for human consumption in Europe, and it's usually served rare in France, followed, predictably, by *Toxoplasma* outbreaks. Budi may not be a carrier, since orangs are mainly vegetarians, but some nonhuman primates in zoos have acquired the bug after eating meat from infected sheep. Perhaps most surprisingly, the pathogen is increasing its range through human-made climate change. With northeastern Europe's warmer, wetter winters, more of the pathogen are surviving, and so are its host species. In fact, *Toxoplasma gondii* may be climate change's oddest bedfellow.

What would cause a rat to find a cat alluring? The slinky sashay? Batonlike whiskers? Crescent-moon-shaped pupils? A stare that nails you in place? Only a foolhardy rodent would cozy up to a cat. Yet rats infected with *Toxoplasma* dramatically change their behavior and find cats arousing. Talk about being in over one's head. There's nothing in it but the briefest frisson for the rat. The cat feeds its belly. But the protozoan zings along the strange trajectory of its life. Since *Toxoplasma* can only reproduce inside a cat's gut, it needs a brilliant strategy to get from rat to cat, and despite its lack of brain power it devised one: hijacking the rat's sex drive. *Toxoplasma*-beguiled rats do feel fear when they smell a cat, but they're also turned on by it, in the ultimate fatal attraction. As with human sexuality, or film noir, a side order of fear isn't necessarily a deterrent.

The cat hunts again, dines on infected prey, and the odd hypnotists thrive. Only cats further the parasite's agenda, but other animals can sometimes ingest the eggs without knowing it and become dead-end hosts. That's why pregnant women are warned not to empty kitty litter or handle cat bedding. Exposure to *Toxoplasma* can derail a fetus, leading to stillbirth or mental illness. Some studies link *Toxoplasma* and schizophrenia. Infected women have a higher risk of suicide than parasite-free women. According to Oxford researchers, it can doom children to hyperactivity and lower IQs. And, for some reason, pregnant women infected with *Toxoplasma* give birth to more boys.

But these new rat–cat findings are only the beginning of an Orwellian saga steeped in irony and intrigue. Worldwide, scientists are posing questions both eye-opening and creepy. If *Toxoplasma* can enslave the minds of rats—animals often studied to test drugs for humans—can it also *alter* the personality of humans? What if that yen to go rock-climbing or change jobs isn't a personal longing at all, robust and poignant as it may feel, but the mischief of an alien life form ghosting through your brain? Is *Toxoplasma* to blame for a hothead's road rage? How about a presidential hopeful's indiscreet liaisons, or a reckless decision made by a head of state? Could a lone parasite change the course of human history?

So when is a whim not a whim? It feels like we have free will, but is a tiny puppeteer pulling the strings of billions of people? For the longest time philosophers, theologians, and college students debated such questions, then neuroscientists joined the fray, and now a body of parasitologists.

When Jaroslav Flegr, of Charles University in Prague, surveyed people infected with *Toxoplasma*, he found clear trends and surprising gender differences. The women spent more money on clothes and makeup and were more flirtatious and promiscuous. The men ignored rules, picked fights, dabbled in risk, and were nagged by jealousy. Both sexes got into more than twice the average number of traffic accidents—as a result of either impulsivity or slowed reaction time.

Rats have proclivities and tastes. Humans have those in spades, as well as sentiments and reveries. But mindset doesn't matter.

All warm-blooded mammals respond to thrill, anticipation, and reward—especially if that includes a wallop of pleasure. Many of the odd behavioral changes scientists attribute to *Toxoplasma* tap the brain's dopamine system, and that's what *Toxoplasma* zeroes in on, rewiring networks to favor its own offspring, even if that means death for the host. Cocaine and other euphoriants use the same dopamine system. As the Stanford neuroscientist Robert Sapolsky explains, "the *Toxoplasma* genome has the mammalian gene for making the stuff. Fantastic as it sounds, a humble microbe is fluent in the dopamine reward system of higher mammals.

"This is a protozoan parasite that knows more about the neurobiology of anxiety and fear than twenty-five thousand neuroscientists standing on each other's shoulders," Sapolsky adds, "and this is not a rare pattern. Look at the rabies virus; rabies knows more about aggression than we neuroscientists do. . . . It knows how to make you want to bite someone, and that saliva of yours contains rabies virus particles, passed on to another person." It's an extraordinary genetic tool for a witless one-celled creature to wield.

Marine mammals and birds are spreading the parasite via water currents and ribbons of air. How many of us may already be unwilling hosts? According to the Centers for Disease Control and Prevention, 10 to 11 percent of healthy adults in the United States tested positive for *Toxoplasma*, and the true figure (most people haven't been tested) is thought to be 25 percent of adults. Some scientists estimate that in Britain, a decidedly cat-loving country, half the population has been infected, in France and Germany 80 to 90 percent, and in countries that favor undercooked meat even more, with nearly everyone an unwitting mark—destiny's child, to be sure, but also *Toxoplasma*'s zombie.

According to Nicky Boulter, an infectious disease researcher at Sydney University of Technology, eight million Australians are infected, and "infected men have lower IQs, achieve a lower level of education, and have shorter attention spans. They are also more likely to break rules and take risks, be more independent, more antisocial, suspicious, jealous, and morose, and are deemed less attractive to women.

"On the other hand, infected women tend to be more outgoing, friendly, more promiscuous, and are considered more attractive to men compared with noninfected controls. In short, it can make men behave like alley cats and women behave like sex kittens."

What does it take to slant an opinion? Advertising, group pressure, financial gain, a charismatic leader? How about a real lowlife, a wheeler-dealer who delights in messing with your mind and harbors primitive drives? Enter the saboteur skillful enough to slowly and subtly change the personality of whole nations—a humble microbe. Some researchers speculate that between a third and half the people on Earth now have *Toxoplasma* in the brain. And it's only one of the many microbes that call us home. Is it possible that what we chalk up to cultural differences may be different degrees of mass infection by a misguided parasite? Kevin Lafferty, a parasite ecologist with the U.S. Geological Survey, also theorizes that cultural identity, at least "in regard to ego, money, material possessions, work, and rules," may reflect the amount of a parasite in a population's blood.

If you're now eyeing your tabby with raised eyebrows, there's no need to panic. Even invisible dictators can be deposed, and *Toxoplasma* responds well to antibiotics. In any case, would it have a greater influence than family dramas, pharmaceuticals, TV, college, climate, love, epigenetics, and other factors in human behavior? It's probably one spice among many. After all, a slew of elements and events influence us from day to day, changing us in cumulative and immeasurable ways. *Toxoplasma* may be but one, and it doesn't lurk in all cat owners or devourers of steak tartare. It may ring its changes only in the presence of certain other microorganisms. How can you tell the dancer from his dance of microbes?

In the garden, all the plants and animals have their own slew of microbial citizens, some sinister, others helpmeets. That takes some getting used to. It's a big paradigm change, one future generations will understand from childhood and capitalize on. In health and medicine, they'll focus on the human ecosystem, our whole circus of human cells, fungi, bacteria, protozoa, and archaea working together, untidily perhaps, but in concert.

When I was growing up, scientists only grew microbes in small petri dishes in their labs, and all bacteria were nasty. In just a decade, we've begun seeing the big mosaic and we're even starting to think in terms of microbes for improving the planet in precise ways: fixing the health of endangered species with wildlife probiotics, ousting invasive species using certain bacteria, sweetening groundwater that's been tainted by pollutants, cleaning up oil spills with voracious grease-loving microbes, helping agriculture feed more people without fertilizers by employing bacteria that make the crops grow faster and more robustly.

The hope is that, just as with genes in the Human Genome Project, if researchers can identify the core microbes that most humans share, then it will be easier to divine which species contribute to specific complaints. This offers a new frontier for fighting illness, one easier to manipulate than the genome, and safer to barge in on than deeply embedded organs like the heart or liver.

New studies suggest that a single pathogen is rarely enough to trumpet disease, because different microbes form alliances. "The real pathogenic agent is the *collective*," says David Relman, an infectious disease specialist at Stanford University. This has sparked a new way of thinking about illness called "medical ecology," which recognizes the collective as the key to our health. In the past, we thought of all bacteria as bad, a contagion to be banished, a horde of invisible dragons. Ever since the end of World War II, when antibiotics arrived like jingle-clad, ultramodern cleaning products, we've been swept up in antigerm warfare. But in a recent article published in *Archives of General Psychiatry*, the Emory University neuroscientist Charles Raison and his colleagues say there's mounting evidence that our ultraclean, polished-chrome, Lysoled modern world holds the key to today's higher rates of depression, especially among young people. Loss of our ancient bond with microorganisms in gut, skin, food, and soil plays an important role, because without them we're not privy to the good bacteria our immune system once counted on to fend off inflammation. "Since ancient times," Raison says, "benign microorganisms, sometimes referred to as 'old friends,' have taught the immune system how to tolerate other harmless microorganisms, and in the process reduce inflammatory responses that have been linked to most modern illnesses, from cancer to depression." He raises the question of "whether we should encourage measured reexposure to benign environmental microorganisms" on purpose.

A baby is born blameless but not microbe-free. Mom transfers microbes through the umbilical cord and down the birth canal, including *Lactobacillus johnsonii* (a bacterium one expects to find in the gut, not the vagina), a bug essential for digesting milk. I was bottle-fed formula, but breast-milk-fed babies grow stronger immune systems because breast milk, often the first source of nourishment, teems with more than seven hundred species of hubbub-loving, life-enhancing bacteria. Researchers are thinking of cobbling them into infant formula to help ward off asthma, allergies, and such auto-immune triggermen as diabetes, eczema, and multiple sclerosis. Babies pick up other useful bacteria in Mom's dirt-and-crumb-garlanded home and landscape. At least, they should.

Doctors are embracing the idea of personalized medicine based on a patient's uniquely acquired flora and fauna, as revealed in his or her genome, epigenome, and microbiome. No more antibiotics prescribed by the jeroboam on the off chance they might prove useful. Instead, try unleashing enough beneficial bacteria to crowd out the pathogen. No more protecting children from the hefty stash of derring-do white-knight bacteria they need but we've learned to regard as icky.

Patients whose gut flora have been wiped out by certain antibiotics are prey to *Clostridium difficile*, an opportunistic weasel of a bug that causes severe, debilitating diarrhea. Once it has taken up residence, it's miserably hard to expel it and restore the good bacteria. What does seem to help, though it's not an image to dwell on, is fecal transplants from a healthy person—an enema full of bacteria to recolonize a stranger's intestines, join the Darwinian fray, and triumph over the pathogens by acting like sailors on leave.

When Kathy Lammens, a stay-at-home mom with four young children, learned that her nine-year-old daughter's battle with colon disease might lead to a colostomy bag, she began looking for alternative therapies. After much research, she decided on do-it-yourself home fecal transplants, tendering one five days in a row. Twenty-four hours after the first, all of her daughter's symptoms improved. Now Kathy, a robust believer, offers a YouTube video with instructions.

One study has revealed that mice with autism don't host the same gut microbes that mice without autism do, and they seep behavior-altering molecules through the body and brain. But researchers find that dosing the mice with the beneficial bacterium *Bacteroides fragilis* eases the symptoms, and so human trials will follow. Another study discovered that if heart patients don't eat enough protein, the good gut microbe *Eggerthella lenta* will steal some of a patient's dose of digoxin, an important heart stimulant.

Some of global warming's unwelcome guests are tiny winged buccaneers carrying invisible stoles of misery. Mosquitoes in Africa and South America are rambling farther north, injecting dengue fever, malaria, West Nile virus, and yellow fever into parts of the world unfamiliar with such scourges. Perfusing our clothes and bedding with insecticides isn't safe, but the diseases infect hundreds of millions of people each year. So the Michigan State microbiologist Zhiyong Xi has been working on the problem in a novel way, by rearranging microbes. When he noticed that mosquitoes carrying dengue fever and malaria were missing the mosquito-loving bacterium *Wolbachia*, he tried infecting the mosquitoes with a heritable strain of *Wolbachia*, and sure enough, the next generations didn't carry either illness, and the lifesaving trait was passed on to their offspring.

It's intriguing to imagine the role a simple microbe may play in someone's relationships and career, and it reminds us that nothing life ever does is simple, or boring. How many threads weave a fleeting thought, let alone a hankering? It also reminds us of the fierce beauty of Earth's organisms, whatever their size, creatures unimaginably complex, breathtakingly frail and yet sturdy, durable, filled with the self-perpetuating energy we call life. A big brain isn't required to concoct sly, world-changing strategies.

AS I GLANCE out at the yard, I'm charmed by nature's details: the magnolia tree's fuzzy buds fattening up for spring; the melting snow on the lawn that's left hundreds of grass follicles; long arcs of wild raspberry canes covered in their chalky lavender winter mask. But I'm also struck by the everythingness of everything in cahoots with the everythingness of everything else. When I look at my hand now, I scout its fortune-teller's lines, and the long peninsulas of the fingers, each one tipped by a tiny weather system of prints; I see it whole, as one hand. But I also know that only a tenth of what I'm seeing is human cells. The rest is microbes.

When all is said and done, both our parasites and we their inn-keepers are diverse—no one hosts the same reeking and scampering microbial zoo. Our microbes can change either in ratio or in kind at the drop of a cookie or in the splash from a locker-room puddle or through an ardent kiss, and then we have to adapt quickly. So it's possible that some diseases really are inherited, but the genes that bestowed them were bacterial. When you think about it, for a major trait to evolve—something grand like the advent of language or the urge to explore—only one gene has to change on the Y chromosome of one man. That would be enough, over many many generations, to create a predisposition or a trend in an entire culture. It all depends on the highjinks of the maddening microbe.

Maybe this should also remind us how much of a pointillist jigsaw puzzle a *personality* really is. As a friend approaches with a smile, we greet a single person, one idiosyncratic and delightful being who is recognizable—predictable, even, at times. And yet every "I" is really a "we," not one of anything, but countless cells and processes just barely holding each another in equilibrium. Some of those may be invisible persuaders of one sort or another: protozoa, viruses, bacteria, and other hobos. But I like knowing that life on Earth is always stranger and more filigreed than we guess, and that both the life forms we see and those we cannot see are equally vibrant and mysterious.

Where does your life story begin? When does the world start whittling your personality and casting your fate? At birth? In the womb? At the moment of conception, when DNA from your mother and father fuse, shuffling an ancient deck of genetic cards and dealing out traits at random from Mom or Dad? Long before womb-time, it would seem, much farther back, before your parents' courtship, even before their parents', in a crucible of choices, daily dramas, environmental stresses, and upbringing. Our genome is only one part of our saga. The epigenome is another. The birdlike microbes singing in the eaves of the body are yet another. Together, they're offering a greatly enriched view of the terra incognita inside us. In the process, sometimes loud as headlines, but more often silent as the glide of silk over glass, how we relate to our own nature is subtly changing.

CONCLUSION:
WILD HEART, ANTHROPOCENE
MIND (Revisited)

NASA's "Blue Marble" photograph of Earth from space gave us an eye-opening image of the whole planet for the first time. Forty years later, "Black Marble" was equally mind-altering, but in a different way: it introduced us to ourselves. Forging a new geological era, we are an altogether different kind of animal from any the planet has ever known, one able to reinvent itself and its world, and manage to survive, despite more twists and turns in daily life than any creature has ever had to juggle. We inhabit a denser mental whorl than any of our stout-hearted ancestors. We're in the midst of a majestic Information Age, but also an ingenious sustainability revolution, a deluxe 3D revolution in manufacturing, a spine-tingling revolution in thinking about the body, a scary mass extinction of animals, alarming signs of climate change, an uncanny nanotechnology revolution, industrial-strength add-ons to our senses, a biomimicry revolution—among so many other "new normals" that we sling the phrase daily.

We understand ourselves on many more spine-tingling levels: how we're changing the planet, other creatures, and each other. This is not just the Human Age. It's also the age when we began to see, for the first time, the planet's interlaced, jitterbugging ecosystems—on the land, in the air, in the oceans, in society—and unmasked our own ecosystems. We've met many of our makers, the mad molecules.

Thanks to revelations in neuroscience, genetics, and biology we're bringing the life and times of *Homo sapiens sapiens* into a much clearer focus. As the "Me" generation gives way to the "We" generation, we're growing more aware of the ties that bind us—even if we're less relaxed in face-to-face encounters.

We humans have so much in common that we can't seem to speak of comfortably: a genetic code, a niche on a small planet in a vast galaxy in an infinite universe; the underrated luxury of being at the top of our food chain; a familiar range of passions and fears; a mysterious, ill-defined evolution from creatures whose thoughts were like a vapor, and before that bits of chemical and chance so small they pass right through the mind's sieve without its being able to fully grasp them. We have in common, despite our extraordinary powers of invention, subtlety, and know-how, an ability to bore ourselves that is so horrifying we devote much of our short lives to activities designed mainly to make us seem more interesting to ourselves. We have in common a world our senses know voluptuously, from one splayed moment to the next, the wind touching one's chapped lips, a just-forgotten chore, the small unremarkable acts of mercy and heroism parents and lovers perform each day, the collective *sort* of creatures we are, whose qualities embarrass us when we stumble upon them in ourselves, but which we're glad to epitomize in movie stars, sports figures, and politicians—people like Neil Armstrong stepping onto the moon, or Thomas Edison spending the last of his days in Florida trying to make rubber from goldenrod. We have in common a fidgeting, blooming, ever-startling universe, whose complex laws we all obey, whether we're born in Tierra del Fuego or Svalbard.

We're each a sac of chemicals, forged in the sun, that can somehow contemplate itself, even if we don't always know where our pancreas is, and are troubled most days by mundaner matters. When we meet, at parties or on the street, we nonetheless feel like strangers. When we find ourselves alone together in an elevator, it is as if we have been caught at some naughty act; we can't even bring ourselves to meet each other's eyes.

It's time we acknowledged our personality—not just as individuals, but as a species. I once knew a woman who checked into a hotel and, upon entering her room, decided she didn't like the design of the small ornamental finials topping the lamps. She phoned the desk and insisted that they be changed. That may seem like radical pickiness, but our personality as a species includes wide streaks of tinkering and meddling. It's an important part of our character; we're unable to leave anything alone. Let's fess up to being the interfering creatures we are, indefatigably restless, easily bored, and fond of turning everything into amusement, fashion, or toys. We Anthrops can be lumbering, clumsy, and immature. We're also easily distracted, sloppy as a hound dog's kiss, and we hate picking up after ourselves. Without really meaning to, we have nearly emptied the world's pantry, left all the taps running, torn the furniture, strewn our old toys where they're becoming a menace, polluted and spilled and generally messed up our planetary home.

I doubt any one fix will do. We need systemic policy changes that begin at the government level, renewable energy replacing fossil fuels, widespread green building practices, grassroots community and nationwide projects, and individuals doing whatever they can, from composting and recycling to walking to work instead of driving.

In the Industrial Age we found it thrilling to try to master nature everywhere and in every way we could think of. In the Anthropocene, we're engineering ways to help the most vulnerable people adapt, and designing long-term solutions to blunt global warming. Humans are relentless problem-solvers who relish big adventures, and climate change is attracting a wealth of clever minds and unorthodox ideas, as we're revisiting the art of adapting to the environment—a skill that served our ancestors well for millennia, while they fanned out to populate the Earth from equator to ice.

We can survive our rude infancy and grow into responsible, caring adults—without losing our innocence, playfulness, or sense of wonder. But first we need to see ourselves from different angles, in many mirrors, as a very young species, both blessed and cursed by our prowess. Instead of ignoring or plundering nature, we need to refine our natural place in it.

NATURE IS STILL our mother, but she's grown older and less independent. We've grown more self-reliant, and as a result we're beginning to redefine our relationship to her. We still need and cling to her, still find refuge in her flowing skirts, and food at her table. We may not worship Mother Nature, but we love and respect her, are fascinated by her secrets, worry about alienating her, fear her harshest moods, cannot survive without her. As we're becoming acutely aware of just how vulnerable she truly is, we're beginning to see her limits as well as her bounty, and we're trying to grow into the role of loving caregivers.

I'm all for renaming our era the Anthropocene—a legitimate golden spike based on the fossil record—because it highlights the enormity of our impact on the world. We are dreamsmiths and wonder-workers. What a marvel we've become, a species with planetwide powers and breathtaking gifts. That's a feat to recognize and celebrate. It should fill us with pride and astonishment. The name also tells us we are acting on a long, long geological scale. I hope that awareness prompts us to think carefully about our history, our future, the fleeting time we spend on Earth, what we may leave in trust to our children (a full pantry, fresh drinking water, clean air), and how we wish to be remembered. Perhaps we also need to think about the beings we wish to become. What sort of world do we wish to live in, and how do we design that human-made sphere?

Our portrait as individuals will exist, for a while, in books, photographs online, and videos, to be sure. But to know us as a species, far-future humans will need to look to the fossil record of the planet itself. That will tell a tale frozen in ribbons of time. What will it say about us?

We're at a great turning, our own momentous fork in the road, behind us eons of geological history, ahead a mist-laden future, and all around us the wonders and uncertainties of the Human Age.

These days, startling though the thought is, we control our own legacy. We're not passive, we're not helpless. We're earth-movers. We can become Earth-restorers and Earth-guardians. We still have time and talent, and we have a great many choices. As I said at the beginning of this mental caravan, our mistakes are legion, but our imagination is immeasurable.

NATURE {'NĀ-CHƏR} N. The full sum of creation, from the Big Bang to the whole shebang, from the invisibly distant to the invisibly minute, which everyone should pause to celebrate at least once daily, by paying loving attention to such common marvels as spring moving north at thirteen miles a day; afternoon tea and cookies; snow forts; pepper-pot stew; moths with fake eyes on their hind wings; emotions both savage and blessed; pogostick-hopping sparrows; blushing octopuses; scientists bloodhounding the truth; memory's wobbling aspic; the harvest moon rising like slow thunder; tiny tassels of worry on a summer day; the night sky's distant leak of suns; an aging father's voice so husky it could pull a sled; the courtship pantomimes of cardinals whistling in the spring with *what cheer, what cheer, what cheer!* Nature is life homesteading every pore and crevice of Earth, with endless variations on basic biological themes. Ex.: tree frogs with sticky feet, marsupial frogs, poisonous frogs, toe-tapping frogs, frogs that go peep, etc.

Archaic: In previous eras, when humans harbored an us-against-them mentality, nature meant the enemy, and the kingdom of animals didn't include humans (who attributed to other animals all the things about themselves they couldn't stand).

Anthropocene: Nature surrounds, permeates, effervesces in, and includes us. At the end of our days it deranges and disassembles us like old toys banished to the basement. There, once living beings, we return to our nonliving elements, but we still and forever remain a part of nature.

ACKNOWLEDGMENTS

Many thanks to the kind souls featured in this book, who welcomed me so graciously into their work lives: Hod Lipson, Ann and Bryan Clarke, Lawrence Bonassar, Bren Smith, Terry Jordan, Matt Berridge, and others. Continued thanks to my agent, Suzanne Gluck, and my editor, Alane Salierno Mason, for all their encouragement and guidance. I'm grateful to the editors of *Orion* and the *New York Times Sunday Review*, who invited variations on a few of the themes. Heartfelt thanks for their support and friendship to my treasured book group (Peggy, Anna, Jeanne, Charlotte, and Joyce); and to Dava, Whitney, Philip, Oliver, Steve, Chris, Lamar, Rebecca and David, Dan and Caroline; literary assistant Kate, for first reading the manuscript; and Liz, who read the manuscript in its many permutations, expansions, and contractions, until eye-glaze finally set in; and to Paul, the only fiction-writing and oldest living wombat.

NOTES

Apps for Apes

5 Humans have cleared so much forest over the past 75 years that the orangutan population has plummeted by 80 percent. The International Union for Conservation of Nature lists Borneo orangutans as endangered, and Sumatran orangs as critically endangered, with only about ten years left for the entire species. One hundred million acres of Indonesian rainforest vanished during Suharto's reign (1921–2008), and regional timber barons have been plundering the forests even faster since then for mahogany, ebony, teak, and other exotic woods. Then there's palm oil, which is used in rayon viscose and many other products that include the words "palm kernel oil," "palmate," or "palmitate" in their ingredients. Once you start looking, it's startling how many foods, shampoos, toothpaste, soaps, makeup, and other products use palm oil. Orangutan Outreach encourages people to boycott all palm oil–laced products, and dozens of multinational companies (McDonald's, Pepsi, et al.) have agreed, for the sake of the rainforests.

Wild Heart, Anthropocene Mind

9 The term "Anthropocene" was coined by the aquatic ecologist Eugene Stoermer (Emeritus, University of Michigan), who used it at a conference, and Paul Crutzen, who currently works at the Department of Atmospheric Chemistry at the Max Planck Institute in Mainz, Germany; the Scripps Institution of Oceanography at the University of California, San Diego; and Seoul National University in South Korea. Andrew Revkin coined the term "Anthrocene" in his book *Global Warming: Understanding the Forecast* (New York: Abbeville Press, 1992), in which he wrote of it as "a geological age of our own making."

10 "According to the BBC News website": http://www.bbc.co.uk/news/world
    -15391515.

10 "Nature," as E. O. Wilson defines it in *The Creation* (New York: W. W. Norton,
    2006), includes "all on planet Earth that has no need of us and can stand
    alone" (15).

Monkeying with the Weather

40 RinkWatch.org: The website launched on January 8, 2012, and over six hun-
    dred skaters at rinks across the continent began reporting.

41 Extreme weather and climate change: The evidence, already overwhelm-
    ing, continues to mount. When climate scientists in Copenhagen exam-
    ined the tide and hurricane history since 1923, they found an ominous link
    between the fever of the oceans and the number and ferocity of hurricanes.
    Warmer seas provoke higher tides and whip up more violent cyclones. For
    ninety years, triggered by the warming climate, more hurricanes have
    lashed our coastlines and spun fiercer winds. James Hansen of NASA's God-
    dard Institute for Space Studies revealed that from 1951 to 1980 only 1 per-
    cent of the planet was stricken with weather extremes (outlandish heat,
    rain, or drought), but between 1980 and 2012 the figure ballooned to 10 per-
    cent of the planet. At that rate, he explained, in the next decade, extreme
    weather will plague about 17 percent of the world. Because it's important
    that we actually see climate change unfolding around us, and feel how it
    touches us personally, the lively online conservation site 350.org hosted "a
    day of global action" on which people around the world highlighted their
    local evidence of climate change. At daybreak in the Marshall Islands, there
    was a demonstration on a diminishing coral reef. In Dakar, Senegal, people
    marked out the margins of storm surges. In Australia, people hosted a "dry
    creek regatta" showcasing a ruinous drought. In Chamonix, France, climb-
    ers marked where the Alpine glaciers had melted. "Connect the dots" was
    the theme. See 350.org, if you want to participate in inspired feats of envi-
    ronmental activism.

Brainstorming from Equator to Ice

53 In *Earthmasters: The Dawn of the Age of Climate Engineering*, Clive Hamilton
    makes the point that in the United States, climate change used to be a bipar-
    tisan concern, but conservative activists have lumped global warming with
    gun control and abortion rights as part of a scurrilous liberal agenda, not
    something apolitical and innately global but a position the liberals have
    cooked up. Knowing whether someone believes in global warming or not,
    you can safely guess his or her politics. As one Republican meteorologist
    noted, it's become "a bizarre litmus test for conservatism." And so they deny

the science supporting it on political grounds, which makes no sense at all.

A cringe-inducing example of that was when President George W. Bush, at the G8 Summit in 2008, having rejected climate change targets, turned back to his colleagues as he was leaving the closing session, raised a defiant fist, and said light-heartedly, "Good-bye from the world's biggest polluter." http://www.independent.co.uk/news/world/politics/bush-to-g8-goodbye -from-the-worlds-biggest-polluter-863911.html.

A Green Man in a Green Shade

80 Patrick Blanc, *The Vertical Garden: From Nature to the City* (New York: W. W. Norton, 2012), 76.

Is Nature "Natural" Anymore?

112 Bill McKibben, "Nature's independence": Bill McKibben, *The End of Nature* (New York: Random House, 2006), 58.

The Slow-Motion Invaders

128 A python can open its jaws wide as a commodious drawer. But how does it digest something that large? By becoming larger itself. Every time a python eats, its heart, liver, and intestines nearly double in size. Scientists are studying the fatty acids of pythons (which seem to be involved) for potential heart drugs for humans.

Some catastrophes can offer bright sides, even the infestation of pythons in the Everglades. I loved biking on paved roadways through Shark Valley before 2000, when the 'Glades still twitched and thronged with wildlife. But a superfluity of raccoons kept raiding the nests of turtles, birds, and gators, eating their eggs and threatening their future. Burmese pythons happen to love the tang of raccoons, and now, as the raccoon population drops, more turtle, bird, and gator eggs can hatch. (However, that doesn't offset the python's impact on a once-lavish ecosystem.)

Many plants may be going extinct, but we're also gathering together domestic, exotic, and native species in novel ecosystems. See R. J. Hobbs et al., "Novel Ecosystems: Implications for Conservation and Restoration," *Trends in Ecology and Evolution* 24 (2009): 599–605.

134 Plankton got their name from the Greek word for "wandering," because they drift helplessly on the current.

134 food chain: Plankton (plants), at the bottom, are eaten by zooplankton (animals); krill, fish, and other sea creatures eat the zooplankton.

140 "This work . . . this wholesale manufacture of wild birds": Jon Mooallem, *Wild Ones* (New York: Penguin, 2013), 206.

"They Had No Choice"

147 The navy may be phasing out dolphin combatants, but according to a National Resources Defense Council study, the United States has been using unsafe sonar in training exercises off the coast of California that has harmed 2.8 million marine mammals over the past five years. See Brenda Peterson, "Stop U.S. Navy War on Whales," *Huffington Post*, March 14, 2014.

For Love of a Snail

156 W. D. Hartman, "Description of a Partula Supposed to Be New, from the Island Moorea," *Proceedings of the Academy of Natural Sciences of Philadelphia* 32 (1880): 229.

  H. E. Crampton, *Studies on the Variation, Distribution, and Evolution of the Genus* Partula: *The Species Inhabiting Moorea* (Washington, DC: Carnegie Institution, 1932).

  Bryan Clarke, James Murray, and Michael S. Johnson, "The Extinction of Endemic Species by a Program of Biological Control," *Pacific Science* 38, no. 2 (1984).

An (Un)Natural Future of the Senses

173 Jun-Jie Gu et al., "Wing stridulation in a Jurassic katydid (Insecta, Orthoptera) produced low-pitched musical calls to attract females," *Proceedings of the National Academy of Sciences* 109 (2012): 3868–73; published ahead of print February 6, 2012, doi:10.1073/pnas.1118372109.

Weighing in the Nanoscale

182 "Is there a sleigh for my illness?": A Cornell neighbor of mine has just invented a lethal "lint brush" for the blood, a very tiny implantable device that snags and kills cancer cells in the bloodstream, before they can transverse the body. We're constantly minting new metaphors for the brain to use as a mental shortcut. One of today's metaphors sliding into common usage in a similar way is "low-hanging fruit." Words are chosen for inclusion in the *Oxford Junior Dictionary* on the basis of how frequently children use them during the day. A great many nature words, such as "kingfisher," "minnow," "stork," and "leopard," have been removed. Some of the new words added were "analog," "cut and paste," "voicemail," and "blog."

184   GraphExeter: Invented by a team at the University of Exeter, GraphExeter is the lightest, most transparent, and most flexible material ever designed to conduct electricity.

185   "piezoelectrical effect" (literally, "pressing electricity"): using crystals to convert mechanical energy into electricity or vice versa.

Nature, Pixilated

190   "mounted shock warfare": Lynn White Jr., *Medieval Technology and Social Change* (London: Oxford University Press, 1962).

190   "Tinkering with plows and horses": See Jared Diamond, *Guns, Germs, and Steel: The Fates of Human Societies* (New York: W. W. Norton, 1997).

192   "Studies also show that Google is affecting our memory in chilling ways": Four studies led by the Columbia University psychologist Betsy Sparrow.

197   Some days it feels like we're drowning in a twittering bog of information: The Twitter haiku site is but one example. The "twaikus" appear too fast to contemplate, which rather defeats the original purpose of haikus. But at 140 characters they're a great way to let off steam, and they're immensely popular.

197   "At some medical schools": In med schools, virtual cadavers aren't intended to fully replace physical cadavers. McGraw-Hill and many other companies have designed software to use in hospitals, pharmaceutical labs, and Internet courses.

197   Stanford's Anatomage . . . Virtual Dissection Table: www.youtube.com/watch?v=6FFd6VWIPrE.

When Robots Weep, Who Will Comfort Them?

212   "when, by the glimmer of the half-extinguished light": Mary Shelley, *Frankenstein*, chapter 5.

228   On September 14, 2013, the annual Loebner Prize for robots that can pass for human went to a chatbot named Mitsuku. However, it ultimately gave itself away in December with this exchange. Q: "Why am I tired after a long sleep?" A: "The reason is due to my mental model of you as a client."

229   "Can we live inside a house": Technological inventions, such as refrigerators and refrigerated train cars, made frozen food possible, including nutri-

tious out-of-season foods, such as frozen fruits and vegetables. As canning opened up the frontiers and women's home lives, food production became more and more industrialized. With these and other innovations, between WWI and WWII, a massive change in domesticity took place. And after WWII many changes to the domestic kitchen environment (see Dolores Hayden's books).

237 "complexity is free": Explained especially well in Hod Lipson and Melba Kurman, *Fabricated: The New World of 3D Printing* (Indianapolis, IN: Wiley, 2013).

Cyborgs and Chimeras

256 "SyNAPSE" is a backronym (a word chosen and acronym made up to fit it) standing for Systems of Neuromorphic Adaptive Plastic Scalable Electronics.

DNA's Secret Doormen

271 "China starts televising": James Nye, *Mail Online*, January 16, 2014.

271 Zdenko Herceg and Toshikazu Ushijima, eds., *Epigenetics and Cancer, Part B* (San Diego, CA: Academic Press, 2010).

Meet My Maker, the Mad Molecule

302 Home fecal transplants: https://www.youtube.com/watch?v=xLIndT7fuGo.

FURTHER READING

Allen, Robert, ed. *Bulletproof Feathers: How Science Uses Nature's Secrets to Design Cutting-Edge Technology*. Chicago: University of Chicago Press, 2010.

Allenby, Braden. *Reconstructing Earth: Technology and Environment in the Age of Humans*. Washington, DC: Island Press, 2005.

Allenby, Braden R., and Daniel Sarewitz. *The Techno-Human Condition*. Cambridge, MA: MIT Press, 2011.

Alley, Richard B. *Earth: The Operators' Manual*. New York: W. W. Norton, 2011.

Anderson, Walter Truett. *All Connected Now: Life in the First Global Civilization*. Boulder, CO: Westview Press, 2004.

———. *Evolution Isn't What It Used to Be: The Augmented Animal and the Whole Wired World*. New York: W. H. Freeman, 1996.

Anthes, Emily. *Frankenstein's Cat: Cuddling Up to Biotech's Brave New Beasts*. London: Oneworld, 2013.

Balmford, Andrew. *Wild Hope: On the Front Lines of Conservation Success*. Chicago: University of Chicago Press, 2012.

Bates, Marston. *Man in Nature*. 2nd ed. Englewood Cliffs, NJ: Prentice-Hall, 1964.

Bauerlein, Mark, ed. *The Digital Divide: Arguments for and against Facebook, Google, Texting, and the Age of Social Networking.* New York: Tarcher/Putnam, 2011.

Benyus, Janine. *Biomimicry: Innovation Inspired by Nature.* New York: William Morrow, 2002.

Berry, Thomas. *The Dream of the Earth.* San Francisco: Sierra Club Books, 1990.

Blanc, Patrick. *The Vertical Garden: From Nature to the City.* Revised and updated ed. Trans. Gregory Bruhn. New York: W. W. Norton, 2011.

Brand, Stewart. *Whole Earth Discipline: Why Dense Cities, Nuclear Power, Transgenic Crops, Restored Wildlands, and Geoengineering Are Necessary.* New York: Penguin, 2009.

Brockman, John, ed. *Culture: Leading Scientists Explore Societies, Art, Power, and Technology.* New York: Harper Perennial, 2011.

———. Introduction by W. Daniel Hillis. *Is the Internet Changing the Way You Think? The Net's Impact on Our Minds and Future.* New York: Harper Perennial, 2011.

Brooks, Rodney A. *Flesh and Machines: How Robots Will Change Us.* New York: Pantheon, 2002.

Bunce, Michael. *The Countryside Ideal: Anglo-American Images of Landscape.* New York: Routledge, 1994.

Carey, Nessa. *The Epigenetics Revolution: How Modern Biology is Rewriting Our Understanding of Genetics, Disease and Inheritance.* London: Icon Books, 2011.

Carr, Nicholas. *The Shallows: What the Internet is Doing to Our Brains.* New York: W. W. Norton, 2011.

Chaline, Eric. *Fifty Machines That Changed the Course of History.* Buffalo, NY: Firefly, 2012.

Chamovitz, Daniel. *What a Plant Knows: A Field Guide to the Senses.* New York: Farrar, Straus and Giroux, 2012.

Church, George, and Ed Regis. *Regenesis: How Synthetic Biology Will Reinvent Nature and Ourselves.* New York: Basic Books, 2012.

Cipolla, Carlo M. *Before the Industrial Revolution: European Society and Economy, 1000–1700.* 3rd ed. New York: W. W. Norton, 1994.

Clark, Andy. *Natural-Born Cyborgs: Minds, Technologies, and the Future of Human Intelligence*. New York: Oxford University Press, 2003.

Clegg, Brian. *Inflight Science: A Guide to the World from Your Airplane Window*. London: Icon Books, 2011.

Cochran, Gregory, and Henry Harpending. *The 10,000 Year Explosion: How Civilization Accelerated Human Evolution*. New York: Basic Books, 2010.

Cockrall-King, Jennifer. *Food and the City: Urban Agriculture and the New Food Revolution*. Amherst, NY: Prometheus Books, 2012.

Cooper, Jilly. *Animals in War: Valiant Horses, Courageous Dogs, and Other Unsung Animal Heroes*. 1983; rpt. Guilford, CT: Lyons Press, 2002.

Cronon, William, ed. *Uncommon Ground: Toward Reinventing Nature*. New York: W. W. Norton, 1995.

Crosby, Alfred W. *Children of the Sun: A History of Humanity's Unappeasable Appetite for Energy*. New York: W. W. Norton, 2006.

Dake, James. *Field Guide to the Cayuga Lake Region*. Ithaca, NY: Paleontological Research Institution, 2009.

Despommier, Dickson. *The Vertical Farm: Feeding the World in the 21st Century*. Foreword by Majora Carter. New York: Picador, 2011.

Diamond, Jared. *Collapse: How Societies Choose to Fail or Succeed*. New ed. with afterword. New York: Penguin, 2011.

———. *The World until Yesterday: What Can We Learn from Traditional Societies?* New York: Viking, 2013.

Dinerstein, Eric. *The Kingdom of Rarities*. Washington, DC: Island Press, 2013.

Dollens, Dennis. *Digital-Botanic Architecture*. Santa Fe, NM: Lumen, 2005.

Drexler, K. Eric. *Radical Abundance: How a Revolution in Nanotechnology Will Change Civilization*. New York: PublicAffairs, 2013.

Dukes, Paul. *Minutes to Midnight: History and the Anthropocene Era from 1763*. New York: Anthem Press, 2011.

Dunn, Rob R. *The Wild Life of Our Bodies: Predators, Parasites, and Partners That Shape Who We Are Today.* New York: Harper, 2011.

Earle, Sylvia A. *The World Is Blue: How Our Fate and the Ocean's Are One.* Washington, DC: National Geographic Society, 2009.

Edwards, Andres R. *The Sustainability Revolution: Portrait of a Paradigm Shift.* Gabriola Island, BC: New Society Publishers, 2005.

Fagan, Brian. *The Little Ice Age: How Climate Made History, 1300–1850.* New York: Basic Books, 2000.

Flannery, Tim. *Here on Earth: A Natural History of the Planet.* New York: Atlantic Monthly Press, 2011.

———. *Now or Never: Why We Must Act Now to End Climate Change and Create a Sustainable Future.* New York: Atlantic Monthly Press, 2009.

———. *The Weather Makers: How Man Is Changing the Climate and What It Means for Life on Earth.* New York: Grove, 2006.

Forbes, Peter. *The Gecko's Foot: Bio-inspiration—Engineering New Materials from Nature.* New York: W. W. Norton, 2006.

Francis, Richard C. *Epigenetics: The Ultimate Mystery of Inheritance.* New York: W. W. Norton, 2011.

Fraser, Caroline. *Rewilding the World: Dispatches from the Conservation Revolution.* New York: Picador, 2010.

Friedel, Robert. *A Culture of Improvement: Technology and the Western Millennium.* Cambridge, MA: MIT Press, 2010.

Fukuyama, Francis. *Our Posthuman Future: Consequences of the Biotechnology Revolution.* New York: Picador, 2002.

Gissen, David. *Subnature: Architecture's Other Environments.* New York: Princeton Architectural Press, 2009.

———, ed. *Big & Green: Toward Sustainable Architecture in the 21st Century.* New York: Princeton Architectural Press, 2002.

Gore, Al. *The Future: Six Drivers of Global Change.* New York: Random House, 2013.

Gorgolewski, Mark, June Komisar, and Joe Nasr. *Carrot City: Creating Places for Urban Agriculture.* New York: Monacelli Press, 2011.

Haeg, Fritz. *Edible Estates: Attack on the Front Lawn.* 2nd ed. New York: Metropolis Books, 2010.

Hamilton, Clive. *Earthmasters: The Dawn of the Age of Climate Engineering.* New Haven, CT: Yale University Press, 2013.

Hannibal, Mary Ellen. *The Spine of the Continent.* Guilford, CT: Lyons Press, 2012.

Hansen, James. *Storms of My Grandchildren: The Truth about the Coming Climate Catastrophe and Our Last Chance to Save Humanity.* New York: Bloomsbury, 2009.

Harman, Jay. *The Shark's Paintbrush: Biomimicry and How Nature Is Inspiring Innovation.* Ashland, OR: White Cloud Press, 2013.

Hauter, Wenonah. *Foodopoly: The Battle over the Future of Food and Farming in America.* New York: New Press, 2012.

Hayden, Dolores. *Building Suburbia: Green Fields and Urban Growth, 1820–2000.* New York: Vintage, 2004.

———. *The Grand Domestic Revolution.* Cambridge, MA: MIT Press, 1982.

Humes, Edward. *Eco Barons: The Dreamers, Schemers, and Millionaires Who Are Saving Our Planet.* New York: Ecco, 2009.

Hutchins, Ross E. *Nature Invented It First.* New York: Dodd, Mead, 1980.

Jablonka, Eva, and Marion J. Lamb. *Evolution in Four Dimensions: Genetic, Epigenetic, Behavioral, and Symbolic Variation in the History of Life.* Cambridge, MA: MIT Press, 2006.

Jackson, John Brinckerhoff. *Discovering the Vernacular Landscape.* New Haven, CT: Yale University Press, 1984.

James, Sarah, and Torbjörn Lahti. *The Natural Step for Communities: How Cities and Towns Can Change to Sustainable Practices.* Gabriola Island, BC: New Society Publishers, 2004.

Jellicoe, Geoffrey, and Susan Jellicoe. *The Landscape of Man: Shaping the Environment from Prehistory to the Present Day.* 3rd expanded and updated ed. New York: Thames and Hudson, 1995.

Kahn, Peter H., Jr., and Patricia H. Hasbach, eds. *Ecopsychology: Science, Totems, and the Technological Species.* Cambridge, MA: MIT Press, 2012.

Kazez, Jean. *Animalkind: What We Owe to Animals.* Chichester, UK: Wiley-Blackwell, 2010.

Keeney, L. Douglas. *Lights of Mankind: The Earth at Night as Seen from Space.* Guilford, CT: Lyons Press, 2012.

Kintisch, Eli. *Hack the Planet: Science's Best Hope—or Worst Nightmare—for Averting Climate Catastrophe.* Hoboken, NJ: Wiley, 2010.

Klein, Caroline, et al., eds. *Regenerative Infrastructures: Freshkills Park, NYC—Land Art Generator Initiative.* New York: Prestel, 2013.

Klyza, Christopher, ed. *Wilderness Comes Home: Rewilding the Northeast.* Hanover, NH: Middlebury College Press, 2001.

Kolbert, Elizabeth. *Field Notes from a Catastrophe: Man, Nature, and Climate Change.* New York: Bloomsbury, 2006.

Kranzberg, Melvin, and Carroll W. Pursell, Jr., eds. *Technology in Western Civilization,* vol. 1, *The Emergence of Modern Industrial Society, Earliest Times to 1900.* New York: Oxford University Press, 1967.

Kurzweil, Ray. *The Age of Spiritual Machines: When Computers Exceed Human Intelligence.* New York: Penguin, 1999.

Langmuir, Charles H., and Wally Broeker. *How to Build a Habitable Planet: The Story of Earth from the Big Bang to Humankind.* Rev. ed. Princeton, NJ: Princeton University Press, 2012.

Lindsay, Ronald. *Future Bioethics: Overcoming Taboos, Myths, and Dogmas.* Amherst, NY: Prometheus Books, 2008.

Lipson, Hod, and Melba Kurman. *Fabricated: The New World of 3D Printing.* Indianapolis, IN: Wiley, 2013.

Lomberg, Bjørn. *The Skeptical Environmentalist: Measuring the Real State of the World.* Cambridge: Cambridge University Press, 2001.

Louv, Richard. *Last Child in the Woods: Saving Our Children from Nature-Deficit Disorder.* Chapel Hill, NC: Algonquin Books, 2005.

———. *The Nature Principle: Reconnecting with Life in a Virtual Age.* Chapel Hill, NC: Algonquin Books, 2012.

MacKay, David J. C. *Sustainable Energy—without the Hot Air*. Cambridge, UK: UIT Cambridge, 2009.

Macy, Joanna, and Chris Johnstone. *Active Hope: How to Face the Mess We're In without Going Crazy*. Novato, CA: New World Library, 2012.

Marris, Emma. *Rambunctious Garden: Saving Nature in a Post-Wild World*. New York: Bloomsbury, 2011.

Marsa, Linda. *Fevered: Why a Hotter Planet Will Hurt Our Health—and How We Can Save Ourselves*. New York: Rodale, 2013.

Marx, Leo. *The Machine in the Garden: Technology and the Pastoral Ideal in America*. New York: Oxford University Press, 1964.

Mayer-Schönberger, Viktor, and Kenneth Cukier. *Big Data: A Revolution That Will Transform How We Live, Work, and Think*. New York: Houghton Mifflin Harcourt, 2013.

McDonough, William, and Michael Braungart. *Cradle to Cradle: Remaking the Way We Make Things*. New York: North Point Press, 2002.

———. *The Upcycle: Beyond Sustainability—Designing for Abundance*. New York: North Point Press, 2013.

McGilchrist, Iain. *The Master and His Emissary: The Divided Brain and the Making of the Western World*. New Haven, CT: Yale University Press, 2010.

McKibben, Bill. *Earth: Making a Life on a Tough New Planet*. New York: St. Martin's Press, 2011.

———. *The End of Nature*. New York: Random House, 2006.

McLuhan, T. C. *The Way of the Earth: Encounters with Nature in Ancient and Contemporary Thought*. New York: Touchstone, 1994.

McNeill, J. R. *Something New under the Sun: An Environmental History of the Twentieth-Century World*. New York: W. W. Norton, 2001.

Moalem, Sharon, with Jonathan Prince. *Survival of the Sickest: A Medical Maverick Discovers Why We Need Disease*. New York: HarperCollins, 2006.

Mooallem, Jon. *Wild Ones: A Sometimes Dismaying, Weirdly Reassuring Story about Looking at People Looking at Animals in America*. New York: Penguin, 2013.

More, Max, and Natasha Vita-More, eds. *The Transhumanist Reader: Classical and Contemporary Essays on the Science, Technology, and Philosophy of the Human Future.* Chichester, UK: Wiley-Blackwell, 2013.

Morrow, Bradford, and Benjamin Hale, eds. *A Menagerie: Conjunctions,* vol. 61. Annandale-on-Hudson, NY: Bard College, Fall 2013.

Musil, Robert K. *Hope for a Heated Planet: How Americans Are Fighting Global Warming and Building a Better Future.* New Brunswick, NJ: Rutgers University Press, 2009.

Natterson-Horowitz, Barbara, and Kathryn Bowers. *Zoobiquity: What Animals Can Teach Us about Health and the Science of Healing.* New York: Alfred A. Knopf, 2012.

Nye, David E. *American Technological Sublime.* Cambridge, MA: MIT Press, 1994.

Pauli, Lori. *Manufactured Landscapes: The Photographs of Edward Burtynsky.* New Haven, CT: Yale University Press and National Gallery of Canada, 2003.

Peterson, Brenda. *The Sweet Breathing of Plants: Women Writing on the Green World.* New York: North Point Press, 2002.

Pipher, Mary. *The Green Boat: Reviving Ourselves in Our Capsized Culture.* New York: Riverhead, 2013.

Pistorius, Oscar. *Blade Runner.* Rev. ed. London: Virgin, 2012.

Revkin, Andrew. *Global Warming: Understanding the Forecast.* New York, Abbeville Press, 1992.

Ridley, Matt. *The Rational Optimist: How Prosperity Evolves.* New York: Harper, 2011.

Rifkin, Jeremy. *The Biotech Century: Harnessing the Gene and Remaking the World.* New York: Tarcher/Putnam, 1999.

———. *The Third Industrial Revolution: How Lateral Power Is Transforming Energy, the Economy, and the World.* New York: Palgrave Macmillan, 2011.

Rolston, Holmes. *A New Environmental Ethics: The Next Millennium for Life on Earth.* New York: Routledge, 2012.

Rosenzweig, Michael. *Win-Win Ecology: How the Earth's Species Can Survive in the Midst of Human Enterprise.* New York: Oxford University Press, 2003.

Sartore, Joel. *Rare: Portraits of America's Endangered Species*. Washington, DC: National Geographic, 2010.

Savulescu, Julian, and Nick Bostrom, eds. *Human Enhancement*. Oxford: Oxford University Press, 2009.

Schmidt, Eric, and Jared Cohen. *The New Digital Age: Reshaping the Future of People, Nations and Business*. New York: Alfred A. Knopf, 2013.

Seaman, Donna. *In Our Nature: Stories of Wilderness*. Athens, GA: University of Georgia Press, 2002.

Sessions, George, ed. *Deep Ecology for the 21st Century: Readings on the Philosophy and Practice of the New Environmentalism*. Boston: Shambhala, 1995.

Shelley, Mary. *Frankenstein*. 1818; New York: Lancer, 1968.

Siegel, Daniel. *Pocket Guide to Interpersonal Neurobiology: An Integrative Handbook of the Mind*. New York: W. W. Norton, 2012.

Stager, Curt. *Deep Future: The Next 100,000 Years of Life on Earth*. New York: Thomas Dunne, 2012.

Steffen, Alex, ed. Introduction by Bill McKibben. *Worldchanging: A User's Guide for the 21st Century*. Revised and updated ed. New York: Abrams, 2011.

Stevenson, Mark. *An Optimist's Tour of the Future: One Curious Man Sets Out to Answer "What's Next?"* New York: Penguin, 2011.

Stock, Gregory. *Redesigning Humans: Choosing Our Genes, Changing Our Future*. New York: Mariner, 2003.

Thomas, Keith. *Man and the Natural World: Changing Attitudes in England, 1500–1800*. New York: Oxford University Press, 1996.

Thomas, William, ed. *Man's Role in Changing the Face of the Earth: An International Symposium under the Co-chairmanship of Carl O. Sauer, Marston Bates, and Lewis Mumford*. Chicago: University of Chicago Press, 1956.

Thurschwell, Pamela. *Literature, Technology and Magical Thinking, 1880–1920*. Cambridge, UK: Cambridge University Press, 2001.

Tobias, Michael, ed. *Deep Ecology*. Rev. ed. San Marcos, CA: Avant, 1988.

Todd, Kim. *Tinkering with Eden: A Natural History of Exotics in America*. New York: W. W. Norton, 2001.

Trefil, James. *Human Nature: A Blueprint for Managing the Earth—by People, for People*. New York: Henry Holt, 2004.

Turner, Chris. *The Geography of Hope: A Tour of the World We Need*. Toronto: Vintage, 2008.

van Uffelen, Chris. *Façade Greenery: Contemporary Landscaping*. Salenstein, Switzerland: Braun, 2011.

Vogel, Steven. *Cats' Paws and Catapults: Mechanical Worlds of Nature and People*. New York: W. W. Norton, 1998.

Wapner, Paul. *Living through the End of Nature: The Future of American Environmentalism*. Cambridge, MA: MIT Press, 2010.

Wellcome Collection. *A Guide for the Incurably Curious*. Text by Marek Kohn. London: Wellcome Collection, 2012.

———. *Superhuman: Exploring Human Enhancement from 600 BCE to 2050*. Ed. Emily Sargent. London: Wellcome Collection, 2012.

Wells, H. G. *The Island of Dr. Moreau*. 1896. Introduction by Alan Lightman. New York: Bantam Dell, 2005.

White, Lynn, Jr. *Medieval Technology and Social Change*. London: Oxford University Press, 1962.

Williams, Mark, et al., eds. *The Anthropocene: A New Epoch of Geological Time?* Theme issue, *Philosophical Transactions of the Royal Society* 369, no. 1938 (March 13, 2011): 833–1112.

Williams, Terry Tempest. *When Women Were Birds: Fifty-Four Variations on Voice*. New York: Picador, 2013.

Wilson, E. O. *The Creation*. New York: W. W. Norton, 2006.

Wood, Elizabeth A. *Science from Your Airplane Window*. New York: Dover Publications, 1975.

Zalasiewicz, Jan. *The Earth after Us: What Legacy Will Humans Leave in the Rocks?* New York: Oxford University Press, 2009.

————, and Mark Williams. *The Goldilocks Planet: The Four Billion Year Story of Earth's Climate*. New York: Oxford University Press, 2012.

Zuk, Marlene. *Riddled with Life: Friendly Worms, Ladybug Sex, and the Parasites that Make Us Who We Are*. New York: Harcourt, 2007.

INDEX

Page numbers beginning with 313 refer to notes.